Fatigue Detection in Athletes using Sensor Based Application

Fatigue Detection in Athletes using Sensor Based Application

Editors

Dr. Avi Noronha

Dr. Neeraj Milind Shahane

Prof. (Dr.) B. K. Sarkar

Prof. (Dr.) Reena Singh

NAMYA PRESS.com

Published by

Namya Press

India: 213, Vardaan House, 7/28 Ansari Road, Daryaganj, Delhi-110002
Email: namyapress@gmail.com
Website: https://namyapress.com

Edition	:	First Published in 2022 (Namya Press)
Title	:	Fatigue Detection in Athletes using Sensor Based Application
Editors	:	Dr. Avi Noronha, Dr. Neeraj Milind Shahane, Prof. (Dr.) B. K. Sarkar, Prof. (Dr.) Reena Singh
ISBN	:	978-93-5545-086-9
Copyright	:	© Dr. Avi Noronha, Dr. Neeraj Milind Shahane, Prof. (Dr.) B. K. Sarkar, Prof. (Dr.) Reena Singh, 2022 All rights reserved

Preface

It's with great happiness that I would like to acknowledge a great deal of people that get helped me extremely through the entire difficult, challenging, but a rewarding and interesting path towards some sort of Edited Book Without having their help and support, none of this work could have been possible.

Through this acknowledgement, I endeavour to thank every person and element, including my guru that has been instrumental in shaping these Research and compiling them to bring it in the form of a book. Although the research started to build up since day one of the inception of the Glocal University, the actual document of it was endorsed first in 2019 during the five years Academy/Industry evaluation.

Each Chapter has been reviewed and altered time and again with the change in practices, teaching, Process, methodology, Algorithm, Novel Work and community support and other similar cases.

The University has always seen in an upward graph with regard to the achievements, accolades and awards until it saw the adverse effects of the pandemic, where Research and practices had to be altered to meet the needs of Researchers students, staff as well as parents.

It also reflects the views of Guid and Co- Guid highlighted in his book and I remain indebted to him for helping me.

I remain thankful to the staff members of the Glocal University in sharing their perspectives towards writing these Research - Vice Chancellor, Glocal University, and Section Heads, coordinators, teachers, non-teaching staff and office staff.

I appreciate the support of my family in being patient with me and my time schedule which also included late hours at work, only to leave very early again the next morning for the last two decades and more.

All I believe is, no person can be self-made, self-sufficient or can do things single-handedly, especially when Complete the research.

Every single person that I have come across with, during this journey has shaped my personality in some manner or the other, a part of which is probably reflected in this book. I wish to thank each of the above members sincerely and wish the best for them always.

— Editors

Contents

Abstract

In swimming, technical performance is a vital indicator of successful performance; however, during phases of high-intensity training, technique can be negatively influenced by fatigue. Advances in video technology have made it possible to increase coaches' capacity to monitor athletes' technical performances during training. Yet research examining the biomechanical responses that occur with fatigue using video methods, which are applicable and relevant to coaches during training, is scarce. The limited research to date that has examined the biomechanical responses that occur with fatigue has been restricted to examining fatigue during race-like situations and have used technology which is not available for use by coaches during training. Whilst this research indicates that changes occur during races, it does not apply to coaches for use during training. As fatigue, and its management, is a vital component of the training process, identifying a method of monitoring fatigue during training, which is applicable to coaches, is essential for athlete development. This research is therefore intended to address this by exploring the implications of 2-dimensional (2-D) video analysis in the management of fatigue, during training, by competitive swimming coaches.

To explore this, this thesis comprises of four studies. The aim of the first study was to investigate whether 2-D video analysis methods currently used by coaches can measure data validly and reliably. To achieve this, the accuracy, precision, reliability and validity of thirty-five variables (thirty-four technical measures and swim time) were calculated using Dartfish Pro Suite motion analysis software, version 6.0 (Dartfish Ltd, Fribourg, Switzerland) and compared to the smallest worthwhile change. By calculating and comparing these measures, only those technical variables which can be measured with accuracy and precision could be determined objectively. A series of fifteen variables (fourteen technical measures and swim time) were found to be precise, valid and reliable when measured using Dartfish Pro Suite motion analysis software, version 6.0.

Using the previously established technical measures, the aim of the second study was to assess if these measures can determine whether technical changes can occur during a high-intensity training session. To achieve this, a group of seventeen elite breaststroke swimmers completed a standard swim set. This involved repeating maximal effort 100m swims on a limited swim-rest time, designed to induce a fatigued state representative of high-intensity training conditions. To determine whether technical changes as a result of fatigue could be detected, the fourteen kinematic technique variables and swim time (fifteen dependent variables in total) were recorded and analysed using 2-D Dartfish Pro Suite motion analysis software, version 6.0 from video recordings of the first and last 100m swim of each swimmer. In addition, 95% confidence intervals were determined to investigate any commonalities or individual differences among swimmers in changes in technical parameters. It was found that during one high-intensity session, technical changes can occur in a group of elite swimmers. The largest changes were shown in leg glide duration (64.6%), swim time (33.2%), stroke rate (35.3%), stroke length (-29.2%), and average velocity (-10.2%) and were shown to have statistically significant ($p < 0.05$) differences between the first and last 25m of the swim set. These changes were also shown to be common amongst all swimmers and occurred early in the swim set.

To evaluate coaches understanding of fatigue and its management during training, the aim of the third study was to assess coaches' current practices and knowledge regarding fatigue during training. To achieve this, a questionnaire was distributed to over 370 coaches throughout the UK. The questionnaire was separated into multiple sections which assessed: coaches' current understanding of the topic of fatigue; the methods coaches' employ to monitor fatigue during a training session; and the processes used to manage fatigue during the training process. It was found that up to 98% of the coaches consider fatigue, its effect and management important in the development of their swimmers. Despite this, there is a lack of consistency in knowledge and methods used by coaches to monitor this. As a result, coaches are continuing to use traditional methods to monitor their athletes which are quick and reliable, specifically stopwatches (100%) and visual observation (98%).

Due to the predominant use of visual observation to monitor fatigue, and the identification of technical changes with fatigue, the aim of the fourth study was to assess whether coaches could visually identify changes in the previously established technical markers and whether this could be improved through education of fatigue and video analysis methods. To achieve this, two groups of ten competitive swimming coaches observed a series of videos of three swimmers taken pre- and post- training, and were asked to identify any technical factors which they perceived to change. One group underwent an intervention using Dartfish Pro Suite motion analysis software, version 6.0 and underwater analysis to assess whether this improved their ability to visually observe fatigue in elite swimmers. The remaining group of coaches acted as a control group and received no feedback. Following the one hour intervention, the coaches' observations slightly improved, however this improvement was not statistically significant ($p > 0.05$) nor retained after 4 weeks. Although the coaches' perceptions of fatigue during training varied, they did show a keen interest in further training and education on fatigue and 2-D video analysis.

The results from this research indicate that 2-D video analysis is an effective and useful tool, which has practical applications: in monitoring fatigue during a training session; guiding training programmes to maximise training potential; and developing coaches' identification and management of fatigue during training through education programmes.

LAY SUMMARY

To achieve the best swimming performance, the best technical actions and skills are required to be learned and developed through training and feedback provided by a coach. Although the intention of training is to perfect an optimal swimming technique and make it automatic through continual practice, hard training may also put pressure on a swimmer's ability to perform technical actions correctly due to tiredness or fatigue. As fatigue and technique are important components of training, finding a way to observe and highlight them during training, which is useable by coaches during training, is essential for athlete development. This research intended to assess the potential of video cameras for this task and examine their use by coaches during training. To achieve this, this research involved four studies.

The first study found out what aspects of swimming technique can be measured and analysed using video and video analysis software. This work identified that video and video analysis software is capable of measuring fourteen aspects of swimming technique and swim time accurately and effectively.

Using the new measurement methods, the second study assessed whether these variables changed or altered as a swimmer tires, or fatigues, during a hard training session. This work identified that during one hard training session, changes occur in swimming technique in an elite group of swimmers and that video analysis could show this. Certain changes in the technique were also shown to be similar amongst the swimmers.

Due to a lack of any research in this area, the third study investigated what coaches currently know about how their athletes tire or fatigue during training, and if they use any methods to monitor or manage this on a regular basis. This work showed that coaches think the fatigue of their athlete during training is very important, yet the coaches' knowledge of this topic was varied and this was shown in the equipment coaches chose to use. Coaches are currently preferring and continuing to use observation as one of their main tools to monitor their swimmers.

Due to the predominant use of visual observation by coaches when monitoring their swimmers, the final study assessed if coaches could observe the changes in technique due to fatigue, without any technological aid such as underwater cameras, and whether a one hour coaching session could improve this observation or not. Although coaches could observe certain technique factors better than others, their responses to observing changes with fatigue was low and the one hour training session was only able to improve this very slightly.

Overall this research highlighted the limited use of video during training in swimming and the varied understanding of the concept of fatigue by coaches. However, it was a topic which many coaches showed a lot of interest in and a willingness to learn more about. To our knowledge this is one of the first pieces of research to test the potential of simple video analysis in terms of monitoring fatigue during training, in swimming. Therefore, video and video analysis may be useful practical tools which could be used by coaches to monitor their swimmers, as well as aid their own professional development.

1

Introduction

A major factor which can determine success in swimming is the technique that a swimmer uses. Lees (2002; p. 814) described technique as *'the relative position and orientation of body segments'*. In swimming, this requires repetitive, forceful actions of the upper and lower limbs at extreme anatomical positions, while maintaining a stable trunk (Maglischo, 2003). The use of optimal technical actions and postures attempts to reduce the water's resistance to forward motion (drag) and to maximise propulsion within the physiological constraints of a swimmer (Toussaint, 2011). These technical positions and skills are learned and developed through training and feedback provided by the coach. Although the intention of training is to automate and hone an optimal technique through continual practice over time, it may also stress the capacity to perform technical actions correctly. One such stressor, common during high-intensity training phases, is fatigue.

Fatigue is a complex, multi-factorial phenomenon which affects numerous bodily processes and the ability to sustain exercise (Alberty et al., 2009, McKenna et al., 2008). Research into fatigue has been on-going for over a century and this extensive body of work has identified numerous underlying mechanisms, sites and effects of fatigue which can vary between individuals and activities (Ament and Verkerke, 2009, Enoka and Duchateau, 2008, Stone et al., 2007). One of the problems encountered in the study of fatigue, despite the vast amount of research into this topic, is the inconsistency with which the term is used and lack of agreement on a single definition (Ament and Verkerke, 2009, Phillips, 2015). To ensure the definition of fatigue used throughout this thesis is clear, fatigue, in the context of swimming and this thesis, is defined as:

'The inability to sustain maximal swimming velocity' (Alberty et al., 2009; p. 638)

As a result of the numerous definitions of fatigue, there are also a multitude of methods of measurement which exist to monitor fatigue (Enoka and Duchateau, 2008). The methods can be both direct (e.g. the quantification of voluntary and electronically stimulated muscle force production and assessment of low frequency fatigue, Vøllestad (1997)), and indirect (e.g. time to exhaustions, neuromuscular, physiological and perceptual measurements, Ament and Verkerke (2009), Hinckson and Hopkins (2005), Swart et al. (2012)). Unfortunately, direct measurements are predominantly lab-based and applicability to real-world sport and exercise is limited (Phillips, 2015). The use of any method depends on a range of factors including: the type of equipment; the type of research and design used; and the ethical considerations (Phillips, 2015). As fatigue is regarded as an on-going process where the signs and symptoms of fatigue may not always be visible, the effects of fatigue could occur at various stages throughout a training session before a point of failure is reached (Ament and Verkerke, 2009). This could result in the continued repetition of altered technical actions, which may be detrimental to optimal swimming performance. The term 'failure' is used in this thesis to refer to the point that exercise ends and that ultimately occurs with the on-going process of fatigue.

While there is an extensive body of research on the multi-factorial mechanisms of fatigue which can occur during or post-exercise (Ament and Verkerke, 2009, Phillips, 2015), the potential effects on technique in swimming have received much less investigation. Previous research has shown that biomechanical changes are apparent with fatigue in various sports, including sprint running in athletics (Salo and Scarborough, 2006); tackling in rugby (Gabbett, 2008); and technical actions in swimming (Alberty et al., 2008). Those focusing on swimming have identified changes in upper body limb parameters, including measures such as velocity, displacement, angles, and stroking parameters (Deschodt et al., 1996, Suito et al., 2008, Tella et al., 2008, Toussaint et al., 2006). These researchers noted that as swimmers fatigue, they attempt to compensate for reduced propulsive forces by increasing SF, decreasing SL, and altering the stroke coordination to sustain an optimal swimming velocity (Alberty et al., 2005). Most of the research to date in this area has relied on comparisons of stroking parameters during maximum effort, competition-style front-crawl swims; however, key questions remain to be addressed.

Research pertaining to the changes in specific actions and limb positions which make up a technical action and the subsequent impact on swim velocity are scarce. Additionally, many of these studies use protocols which replicate race scenarios of single swims. Swimmers are renowned for covering large distances during every training session and consequently the point at which athletes may develop faulty technique during training, which may serve to add to the effects of fatigue during races, could have been missed (Elbe et al., 2015). Therefore, the present study is designed to redress these factors by assessing the effects of fatigue during a training-like situation.

The identification of fatigue during training is important for several reasons. Firstly, fatigue is seen as an integral part of the training process in terms of athlete adaptation and development. During training, athletes are stressed to a higher level than previously tolerated by using the principles of training and increasing the training load, determined by the intensity, duration and frequency of training sessions (Smith, 2003). This can induce short-term fatigue effects and, following a recovery period, fitness gains can be achieved which results in super-compensation, or adaptations, higher than their usual capacity (Smith, 2003). Training is often designed in a cyclical way to allow time for recovery with progressive overload, often using a method called periodization (Stone et al., 2007). The negative stress resulting from high-intensity training sessions, known as overreaching, is seen as a normal part of training and if the training load and recovery process are balanced correctly, positive adaptations (or super-compensation) and improved performance can follow (Halson and Jeukendrup, 2004). If these factors are not balanced, negative effects of fatigue or poor performance can occur and begin to accumulate over time (Robson-Ansley et al., 2009). The accumulation of the long-term consequences of fatigue can eventually result in overtraining, and is seen as detrimental to performance (Bell and Ingle, 2013, Halson and Jeukendrup, 2004). Both overreaching and overtraining are defined as '*an accumulation of training and/or non-training stress resulting in a decrement in performance capacity with or without related signs and symptoms of overtraining in which restoration of performance capacity may take a certain duration*' (Halson and Jeukendrup, 2004; p. 969). The two conditions are differentiated by the duration of decrement in performance and recovery time (Meeusen et al., 2013). Although overreaching is viewed as a normal part of the training process, it is also perceived to be linked to overtraining; however, this has not been conclusively confirmed (Halson and Jeukendrup, 2004, Meeusen et al., 2013). A number of other factors in addition to training can impact these two conditions, including: the athlete; activities outside the sport; nutrition; and health (Bell and Ingle, 2013, Meeusen et al., 2013, Robson-Ansley et al., 2009).

Secondly, if swimmers are training in a fatigued state, such as during periods of highintensity training, this could result in the repeated performance of incorrect technical actions (Moore and Stevenson, 2010). As part of the purpose of the training process is to perform repetitive exercises designed to induce automation in the execution of technical actions, this could induce the automation of sub-optimal technical actions when a swimmer fatigues, similar to that observed during the latter stages of a race performance (Maglischo, 2003, Smith, 2003).

Thirdly, if swimmers are performing technical actions incorrectly over many repetitions in a training session, this may result in musculo-skeletal adaptations, which are detrimental to performance, such as muscular imbalances, altered muscle structure and function, muscle damage, or postural changes (Kluemper et al., 2006, Thow, 2010). Finally, these factors may increase the stress on other body segments possibly increasing the risk of common swimming injuries (Becker and Havriluk, 2006, Grace, 1985). It is imperative to optimise the training time available to athletes by ensuring it is getting the best out of the athletes and preventing injury. Due to the potential positive and negative effects of fatigue and its fundamental role in the training process, accurate understanding and management are vital. It is essential to monitor and analyse the effects of fatigue during training so that these effects can be recognised.

Technical performance is traditionally studied using the discipline of biomechanics (Bartlett, 2007). Biomechanics is defined as the '*application of mechanical laws to living structures and biological systems, specifically the loco-motor system of the human body*' (Hay, 1993; p. 2). Although there are varying contexts of biomechanics, the present focuses on biomechanics in sport only. This involves the '*study and analysis of human movement patterns in sport*' from a performance enhancement or injury reduction perspective (Bartlett, 2007; p. xvii). Sports Biomechanics is known as the science underlying sporting techniques and provides a basis on which to evaluate the various techniques that are used in swimming, as well as any effects which may be a result of fatigue (Hay, 1993). There are two sub-branches of biomechanics, kinematics and kinetics. Kinematics is known as the study and description of bodies in motion and often deals with the observable features of movement (Robertson et al., 2013). Kinetics attempts to understand why motions occur or the underlying forces involved in motion (Robertson et al., 2013). Although both branches of biomechanics are important in technique analysis and development, kinetics and its use in an applied sporting environment has been scarce and difficult to implement due to the complex and mathematical nature of this subdivision. Kinematic analysis, on the other hand, is within the capacity of coaches to observe, interpret, and implement strategies relating to training and technique development based on their observation. As kinematics is more universal to all the individuals involved in the monitoring, analysis and feedback of technical performance to athletes, and due to the scarcity of research on the effects of fatigue on technique in swimming, this thesis focuses on kinematic biomechanics only. By using the kinematic approach, this thesis will not explain why any changes in motion or the underlying forces occur due to fatigue; however, due to the novelty of this research topic, data from this thesis will be able to provide a basis on which to evaluate the influence of fatigue on technique in swimming during training. This is a vital step in the research process as it can aid in the identification and monitoring of the effects of fatigue during training which is essential for coaches and athlete development.

The monitoring, analysis and feedback of technical performance is completed by two main individuals; the sport biomechanist (scientist), and the sport coach. Although each individual plays a vital role in the development of an athlete and they have the shared focus and intent of improving performance, their approach to this dilemma differs greatly in terms of monitoring and analysis (Robertson et al., 2013). Sports biomechanists tend to predominantly use a quantitative approach. This approach describes and analyses movement numerically, is very data driven and provides large quantities of information (Bartlett, 2007). It mainly uses methods such as video or automatic marker tracking systems, with additional methods including electromyography, force and pressure measures, as well as statistical modelling and computer simulation (Bartlett, 2007). Coaches on the other hand tend to use a qualitative approach. In this approach performances are analysed descriptively by interpreting movement patterns, and it uses a structured, multi-disciplinary approach (Bartlett, 2007). Qualitative analysis mainly involves the use of methods such as video recordings or observations, and software packages such as Silicon Coach (Silicon Coach, Dunedin, New Zealand) or Dartfish (Dartfish Ltd, Fribourg, Switzerland) (Kerwin and Irwin, 2008, Robertson et al., 2013). This method is seen as being inexpensive, systematic and coachfriendly; however, it does require considerable knowledge of technique, depends on the subjective opinion of the observer and may be unreliable (Bartlett, 2007). As a result, the subjective nature of qualitative analysis makes it restricted to the coaches' analysis of actions that can last fractions of a second (Fleming et al., 2010). As the most common method to monitor technique from both approaches is video imaging or motion capture, this method is predominant in this thesis.

Prior research has also indicated that a gap exists between theories, research and its application to practice (Bishop, 2008a). As researchers tend to utilise complex quantitative methods, these are often expensive, require technical skills, and entail long data collection and analysis phases which can delay the feedback to both the coach and swimmer (Liebermann et al., 2002). As a result, these methods are often far removed from the coaching environment and cannot be applied directly to the coaching section (Kerwin and Irwin, 2008). Many coaches are unable to co-ordinate the provision of expert services and support of their athletes, nor use the same equipment or measures in a training environment (Bishop, 2008a, Kerwin and Irwin, 2008). Despite improvements in video technology, coaches are continuing to use a qualitative approach and there continues to be an inconsistency between the methods used by coaches and sport scientists to monitor and assess athletes' performance during training. There is also a disparity in the use of biomechanical knowledge or principles in the monitoring of fatigue of swimmers. Addressing these divergences in communication, research, equipment and the impact of fatigue are essential if athlete development is to continue through optimal training conditions.

1.1. STATEMENT OF PROBLEM

Swimming is a sport which is highly technique dependent, and involves high-intensity, prolonged training sessions (Maglischo, 2003). During these sessions, fatigue may influence the ability to perform technical skills. To ensure any effects are not detrimental to performance, swimmers rely on the feedback from coaches to develop and improve their ability to cope with, and manage, fatigue during training and maintain an optimal swimming technique. Coaches currently rely on qualitative methods of analysis (Kerwin and Irwin, 2008), with no research pertaining to their knowledge or understanding of fatigue and its influence on technique and biomechanics. Due to the major role technique plays in swimming performance, and the importance of coach feedback in its development, further research is required to understand and enhance coaches' current practices regarding the monitoring of fatigue during training. This will involve understanding the effects of fatigue and the current practices of coaches using methods which are available to coaches for use on a regular basis. There has been a small body of work that has examined the effects of fatigue in swimming (Alberty et al., 2005, Conceição et al., 2014, Suito et al., 2008, Toussaint et al., 2006) but this body of work has used methods unavailable to coaches, and focused upon the athlete. This work has yet to provide an effective and appropriate way to monitor fatigue during training which is applicable and relevant to coaches and the training environment.

1.2. AIMS OF RESEARCH

The purpose of this research was to explore the implication of video analysis methods available to elite competitive swimming coaches to aid the monitoring of fatigue within training. It is therefore intended to address the following research aims:

1. Establish the quality of 2-dimensional (2-D) kinematic technical measures using Dartfish video analysis software, version 6.0, for a breaststroke swimming action and underwater analysis.
2. Investigate a series of 2-D technical kinematic markers as indicators of acute fatigue.
3. Examine coaches' current knowledge and practices regarding monitoring fatigue during training.
4. Determine whether the use of technical indicators of fatigue can aid coaches' abilities to observe and identify changes in technique which occur as a result of fatigue.
5. This thesis concludes with a summary of the results, implications, recommendations and suggestions for future use. In addition, at the end of each chapter there is a summary box for that specific chapter, as below.

2

Literature Review

In Chapter 1, fatigue was introduced as an important aspect to be considered in the training and coaching process in swimming. Throughout this thesis, the coaching process will be defined as "*the purposeful improvement of competition sports performance, achieved through a planned programme of preparation and competition*" (Lyle, 1999; p. 8). Despite the influence of fatigue on swimming performance and the role feedback from coaches' plays in technique development; little is known about the influence of fatigue on technical performance during training or current coaching practices to monitor it. Prior to conducting research to address this gap in knowledge, it is necessary to analyse and evaluate the applicable literature to provide a foundation for the research. In the present chapter, three pertinent areas of extant research will be reviewed: technique in swimming and the implications for training; the effects of fatigue on technical performance; and the role of video analysis in providing feedback on technique.

2.1. TECHNIQUE IN SWIMMING AND THE IMPLICATIONS FOR TRAINING

The main factor in swimming which determines race time is the mid-pool swimming speed (Mason and Cossor, 2000, Pai et al., 1984, Thompson et al., 2000). This indicates that understanding and practice of mid-pool swimming technique is imperative for breaststroke swimmers (Mason and Cossor, 2000). Throughout the training process, swimmers prepare for the demands of competition by continually developing and honing the technical performance of their stroke to complete a race distance (in accordance with the governing rules) in the fastest time possible (Hay, 1993).

The stroking time is determined by the average swim speed over a set distance. A number of biomechanical models have been developed to identify the factors that determine performance (Hay, 1993). These models all highlight the importance of three key areas for swimming speed, namely; the stroking parameters stroke length – SL, and stroke frequency – SF; hydrodynamic propulsive and resistive forces; and the kinematic technical actions a swimmer uses (see Figure 2.1.).

To maximise a swimmer's ability to perform technical skills effectively in swimming, it is important that the coach: understands the factors which can influence swimming speed and these relationships; can observe and identify particular deviations in a swimming stroke; and knows whether these deviations are mechanical or due to the influence of other factors (Persyn et al., 1983). Further to this, the coach must know which factors can be altered by training and how to do so (Persyn et al., 1983). As an understanding of these areas is required for a coach to interpret any effects of fatigue on technique, the relationship of these areas are subsequently briefly discussed.

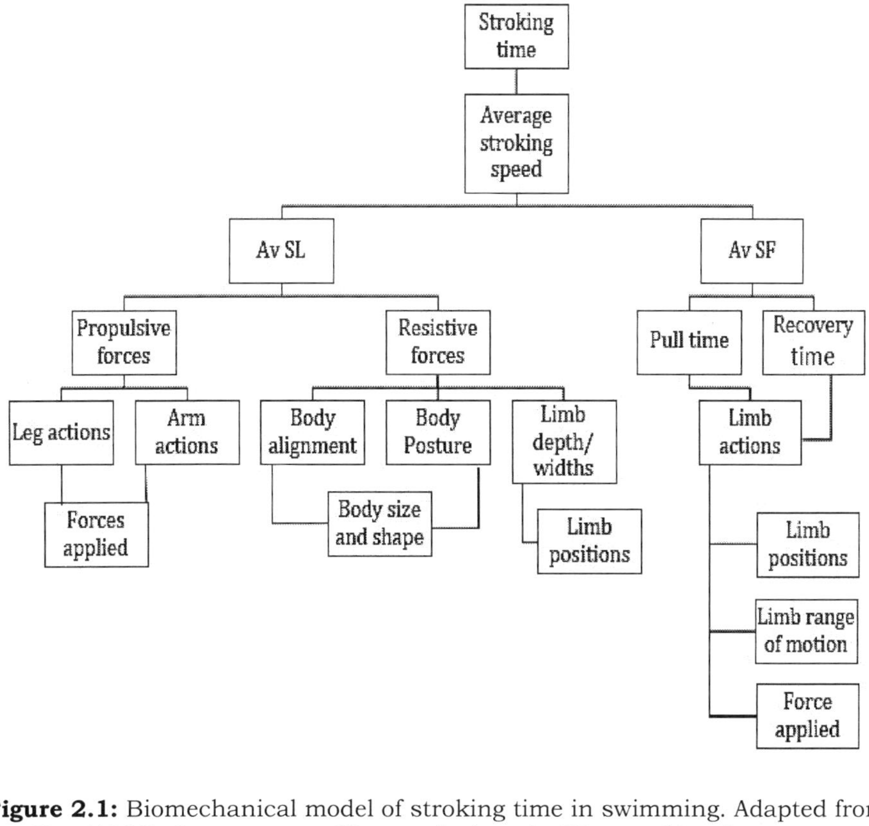

Figure 2.1: Biomechanical model of stroking time in swimming. Adapted from Hay (1993). Av = average; SL = stroke length; SF = stroking frequency.

2.1.1. Stroking Parameters

A swimmer's speed is the product of the two stroking parameters, SL and SF (Hay, 1993, Maglischo, 2003). SL is defined as *'the distance per stroke' (metres)'*. SF is defined as *'the number of stroke cycles swimmers take each minute (cycles/min)'*. It is reported that a combination of SL and SF determine successful swimming performance and as a result a large number of studies in swimming have analysed the relationships of these two parameters since the pioneering work of East in the 1970's (Craig and Pendergast, 1979, Craig et al., 1985). These studies primarily identified that SL and SF have an indirect association in that if one increases the other decreases and vice versa. For example, to increase SF, swimmers tend to reduce the time they spend pulling and this usually leads to a reduction in SL. These studies have also highlighted a number of aspects which affect stroking parameters and their relationship to swimming speed, including: the stroke (Hellard et al., 2008), gender (Takagi et al., 2004), ability/skill level (Chollet et al., 1997, Takagi et al., 2004), physique characteristics (Kennedy et al., 1990), and race performances (Mason and Cossor, 2000, Takagi et al., 2004, Thompson et al., 2004).

These pieces of research emphasise that:

SF and SL are seen as indicators of motor processes (Alberty et al., 2008).

SL is seen as the most important predictor of swim performance in front-crawl, back-crawl, and butterfly and can discriminate levels of expertise due to its capacity to indicate efficiency (Chollet et al., 1997, Craig et al., 1985, Pai et al., 1984). It is suggested that the higher the SL, the higher the economy as reflected by an ability to reduce drag, create force and, as a result, high propulsive efficiency (Maglischo, 2003).

Alternatively, SF has been suggested to be the most discriminating factor in breaststroke swimming (Seifert and Chollet, 2005).

Swimmers often compensate for a decrease in SL with an increase in SF during race performances (Alberty et al., 2008). Breaststroke is seen as having more of a decline in SL than any other stroke (Alberty et al., 2008, Craig and Pendergast, 1979, Thompson et al., 2004).

Expert swimmers maintain their speed better than less skilled swimmers (Chollet et al., 1997) by manipulating the decrease in SL and the increase in SF during race performances (Alberty et al., 2008, Craig et al., 1985, Nikodelis et al., 2005).

To increase velocity over short distance events, swimmers will increase their SF. Swimmers have been shown to gradually increase their SL over race distances of 50-200m, yet SL decreases thereafter in all swimming strokes (Craig et al., 1985).

Differences between swimmers' SL is related to anthropometric differences. SL is statistically significantly longer in men than women. For example, taller swimmers typically use a slower SF and cover more distance with each stroke than shorter swimmers (Maglischo, 2003).

Literature has also shown that two swimmers performing at a similar swimming speed do not necessarily have the same SL-SF combination (Chatard et al., 2003). When swimmers increase one of these two factors they must ensure that the other does not suffer a comparable decrease and find the optimal individual combination of these two parameters to reach and maintain the highest possible speed (Sidney et al., 1999). This may be the reason for some of the inconsistencies in SL and SF literature. These studies highlight and emphasise the individual nature of SL and SF and, due to their importance in obtaining and maintaining the highest possible speed, the necessity to ensure that the optimal combination of SL and SF is achieved.

Stroking parameters play a vital role in race performance and have become wellestablished in biomechanical analysis in swimming, yet there has been very little research to assess these variables during training scenarios in which they would be learned and developed. Wakayoshi et al. (1996) reported that an adaptation of endurance training is the ability to maintain a long SL in races and identified speed increases associated with SL during a 400m swim after six months of training, with no change in SF. Unfortunately, the details of the type of training undertaken were not described in depth, nor were swimmers analysed individually. Further research investigating the effects of different types of training formats, intensities and durations on individual swimmers, is needed to fully understand the implications of training on these parameters.

In addition, literature pertaining to understanding how coaches utilise these parameters to guide technique development or monitor their swimmers during training is also scarce. Alberty et al. (2008) suggested that the previous research on SL and SF has led coaches to develop training methods where only limited changes in SF or SL are permitted and strategies have to be developed to maintain a high speed. This has resulted in the design of training sets in which speed and the number of stroke cycles permitted per distance unit is fixed or the same SF-SL combination, assuming it is sufficient to stabilise stroke technique (Alberty et al., 2008, Alberty et al., 2011). More specifically coaches can require the swimmers to change their usual SF-SL combination at a given speed. For example, a given speed has to be maintained with a lower SF than one naturally adopted at this pace (Alberty et al., 2011). To ensure swimmers are learning and developing these skills optimally during training, further research is required in 'training-like' settings to further understand how these parameters are affected during training.

During training and competition, a range of methods are often employed to quantify SF and SL. These include using predefined variables, such as SL or speed, using stopwatches specifically designed to measure SF, or counting the number of strokes performed during a set distance and the time taken to complete that distance (Chollet et al., 1997, Seifert et al., 2007). The data obtained from these methods however, is questionable. Chollet and Pelayo (1999) confirmed this by reporting statistically significant differences in the use of a range of calculations of SL. They reported that when measuring SL from average speed and SF, values were higher than that for speed calculations not accounting for starts and turns. Further, when SL was calculated by dividing the distance swum by the number of stroke cycles, these errors increased. Although these methods are used daily by coaches in a practical setting, any comparisons with previous literature must be interpreted with caution if alternative methods are used.

During particular phases of training, swimmers are subject to long periods of highintensity, fast swimming in which they are required to sustain high swimming speed. Research during races has shown that the higher the SF values during sprinting, the higher the energy expenditure and thus the capacity to sustain this for long durations is reduced (Chollet et al., 1997, Craig et al., 1985, Maglischo, 2003). Ensuring that swimmers are practising and maintaining the most stable stroke technique in preparation for the latter stages of a race performance appears to be an important goal of high-intensity training, yet little is known regarding the methods coaches employ to monitor technique during training. Understanding the implications of training on SL-SF combinations is therefore an important aspect to maximise a swimmer's technical capacity in swimming and assess whether the strategies swimmers are employing are effective.

2.1.2. Propulsive and Resistive Forces

As shown in Figure 2.1, the SF and SL of a stroke cycle are governed by the ability of a swimmer to influence the interaction of two important forces; propulsion and water resistance (Toussaint and Beek, 1992).

Although the present thesis is not directly measuring propulsion or water resistance, an understanding of these concepts is needed to appreciate the relationship between kinematic technique variables, stroking parameters and overall swimming speed. Thus propulsive and resistive forces will be briefly introduced and explained.

Successful performance is determined by the propulsive forces a swimmer generates during a stroke cycle whilst minimising the resistive forces acting on the body (Caty et al., 2007). Propulsion is the force applied to propel the body forward through the water, achieved by co-ordinating the limbs, trunk and head (Caty et al., 2007, Vorontsov and Rumyantsev, 2000). Studies analysing intra-cyclic velocity variations initially highlighted that breaststroke comprises four propulsive phases; two leg propulsive phases (at the end of the leg extension and the leg in-sweep) and two arm propulsive phases (the hand in-sweep and the hand out-sweep). As with the other three swimming strokes, the highest propulsive forces and forward acceleration are achieved using the arms when swimming at race-pace (Leblanc et al., 2007). Contrary to the other swimming strokes, the leg kick action in breaststroke is also considered to be a dominant propulsive force and has a much larger contribution to the total propulsive force in this stroke than in front-crawl and back-crawl (Mason and Cossor, 2000).

Water resistance, or active drag, is the force which resists swimmers' forward motion and movement through water. Active drag is comprised of three contributions: form, surface and wave drag (Toussaint, 2011). Form drag is the water resistance relating to the space, shape and position of the body relative to the oncoming flow of water (Zamparo et al., 2009). The magnitude of form drag is governed by the speed at which the swimmer is travelling forward through the water and by the cross-sectional area they present to the oncoming flow (Toussaint, 2011). Surface (or friction) drag is the drag created by the friction between the body of a swimmer and the water particles (Zamparo et al., 2009). Wave drag is regarded as the turbulence created at the water surface which acts to inhibit forward motion (Vorontsov and Rumyantsev, 2000). The magnitude of wave drag is proportional to the swimming speed of the swimmer, with higher swim speeds creating greater drag effects and reducing the forward motion of the swimmer (Maglischo, 2003).

The type of drag which has the most effect has been debated for several decades with Vorontsov and Rumyantsev (2000) determining that wave and form drag increase with the cube and square of the swimming speed, respectively, and suggesting that surface drag is often regarded as negligible in comparison. Clarys (1979) also concluded this but Sharp et al. (1988) concluded that surface drag was not negligible, indicating more research is needed in this area. As this study is focusing upon technique factors and the surface of the swimmer is not being considered, this study will focus on the effects of technique variables and reducing wave and form drag.

According to kinematic analysis and concepts of drag, any movements or actions which increase the frontal area or create turbulence in the water are likely to increase drag or limit the effectiveness of propulsion and, as a result, will have detrimental consequences on a swimmer's speed (Zamparo et al., 2009). This has been noted in comparisons of elite and non-elite swimmers, with elite swimmers possessing longer SL and more economical stroke patterns and body movements and, as a result, having a lower active drag (Nikodelis et al., 2005). As both the maximisation of propulsion and minimisation of drag are dependent upon the swimmer's speed, body shape, streamlining and limb actions or movements in the water (Zamparo et al., 2009), kinematic aspects of technical performance are highly relevant for stroking parameters, swimming efficiency and overall swimming velocity.

2.1.3. Kinematic Technique Actions

A swimmer's goal is to maximise swimming speed by maximising the SL-SF relationship through the optimisation of propulsion and minimisation of resistance and this is achieved by obtaining a streamlined body shape and co-ordinated efficient technical movement pattern (Seifert et al., 2005, Seifert et al., 2010, Toussaint and Beek, 1992). However, due to water being one hundred times denser than air, applying forces is more difficult than on land and any movement in an aquatic medium is greatly affected by water resistance. Although stroking parameters are commonly and frequently used in training, Seifert et al. (2004b; p. 658) stated that '*they are not sufficient in themselves to examine a swim race, because they do not provide technique or coordination measures*'. The balance of the resistive and propulsive forces in swimming are caused by the limbs and trunk obtaining an optimal balance (Persyn et al., 1983) and thus it is the analysis of kinematic variables which would enable a more comprehensive understanding of the influential aspects of technique on swim speed and is one of the main points of interest in biomechanics in swimming (East, 1970, Seifert et al., 2004a).

Breaststroke is the slowest of the four competitive swimming strokes (Maglischo, 2003) due to the underwater arm and leg recovery actions which cause large amounts of drag (Leblanc et al., 2009). This can cause swimmers to lose considerable momentum during these stroke phases (Maglischo, 2003). In breaststroke the changes in the positions of the swimmer's limbs probably have a greater influence on the resistance encountered than corresponding changes in other strokes. It is imperative that stroke technique and mechanics are performed correctly to maximise the propulsion achieved in this stroke (Craig et al., 1988) and minimise the drag. Due to these factors, this thesis will focus on the breaststroke technique. The effect that various body positions and limb movements have on minimising resistance and maximising propulsion has been studied by a number of people, some specifically on breaststroke swimming and some relevant due to the similarities of the four swimming strokes, and these will now be discussed briefly. Additional literature pertaining to research on the breaststroke technique is provided in Chapter 4.

Since the late 1970's propulsive forces have been strongly linked to kinematic arm and hand parameters (Berger et al., 1995, Schleihauf, 1979) and it is widely acknowledged that swim speed and forward propulsion of the body is partly explained by the horizontal and vertical arm actions and positions during the stroke (Deschodt et al., 1999, Deschodt et al., 1996, Schleihauf, 1979). A number of studies in front-crawl have established a link with the backward displacement of the wrist, elbow and shoulder arm segments to increasing swimming speed (Deschodt et al., 1996, Vorontsov and Rumyantsev, 2000). Differences in these displacements were shown between skilled and less skilled swimmers, with less skilled swimmers demonstrating a larger backward displacement of the arms (0.4-0.5m compared to 0.6-0.7m) (Vorontsov and Rumyantsev, 2000). Skilled swimmers have also been shown to have large vertical motions of the arm segments which also relates to the swimmer's forward velocity (Deschodt et al., 1996). These studies identified that hand depth can range between 0.4-0.8m and recommend this depth to establish a backward facing position of the hand at the catch phase (Costill et al., 1992, Deschodt et al., 1996). Cappaert (1999) speculated that the vertical displacement of the arm is related to the elbow angle, thus any changes in the elbow angle would result in a deeper or shallower pull pattern which could influence the propulsion or drag generated by these actions.

Previous findings in the literature suggests swimmers tend to exhibit a 90° elbow angle during the underwater arm phases; however, observations indicate that the elbow follows a flatter, lateral pathway than the wrist during the underwater pull phase and there are great individual differences in the lateral deviations of the arm underwater (Deschodt et al., 1996). This is important due to the influences the elbow angle has on the arm trajectory during the underwater phase, the stroke efficiency and the power of the arm-pull (Cappaert, 1999,

Vorontsov and Rumyantsev, 2000). Aspects such as a greater vertical or lateral trajectory of the hand is thought to increase the frontal area of propulsive segments and allow swimmers to exert an increased force, both in an upwards and backwards motion (Maglischo, 2003). This is believed advantageous as once swimmers start moving water, they cannot elicit the same reaction force and thus Maglischo (2003) proposed that swimmers who stroked laterally to continuously find still water and achieve a greater reaction force than when pulling directly back (Maglischo, 2003). In addition, literature on hand and arm actions has revealed that even arm motions that are non-propulsive are still believed to be essential in terms of enhancing the propulsive aspects of the stroke cycle and increasing the SL to increase swimming speed (Maglischo, 2003). Therefore the kinematic parameters of the arms have large implications for swimming speed and it is imperative to ensure that these are being completed correctly. Unfortunately, kinematic arm and hand parameters are often measured using 3-D video analysis, which is not often applied into a coaching environment. It has also not been investigated whether coaches can observe hand and arm actions which take place under the water or whether video aids are needed to ensure that technical feedback from coaches to swimmers is effective.

The impact of slight changes in limb positions was also noted by Counsilman (Hay, 1993) who showed that when towed in a prone position, changes in head position could influence the drag experienced by the swimmer. This was further investigated by Zaidi et al. (2008) using computational fluid dynamics. They also identified that an optimal position of the head can influence the drag forces experienced by a swimmer. The findings of these studies indicated that when the head is held in a high position there is a significant increase in the drag related to an increase in cross-sectional area due to legs and feet 'dropping' as the head is raised. Similar findings have been established in terms of the angle of the trunk (Zamparo et al., 2009), finger spacing (Minetti et al., 2009, Remmonds and Bartlett, 1981) and thumb positioning (Marinho et al., 2009). Unfortunately, these studies are based on race performances and the outcomes of computer simulations and there have been no investigations into how coaches educate swimmers or observe the specific kinematic technical actions which can impact the drag and propulsion experienced by a swimmer, and therefore their overall swimming speed.

Finally, for many years the role of the leg kick in swimming has been well debated, and for the other three strokes its contribution to propulsion is thought to be minimal and to be used primarily for stabilisation of posture and drag reduction. It has been noted in front-crawl and back-stroke that certain kicking depths can influence the drag experienced by swimmers, again measured during towing (Alley, 1952, Kruchosky, 1954). In breaststroke however, the kick is considered to play a much larger role in forward propulsion than the other strokes. Despite this, there has been very little research in this area.

As with the whole stroke there are several different kick styles compared. These comparisons have established that the whip kick is the superior kick in terms of speed, propulsive force, economy of movement and tempo (Counsilman, 1948). It is characterised by either its 'forceful backward push' or 'outward and down propeller-like' action (Maglischo, 1982). One specific factor which has arisen from research into the breaststroke kick is the range of styles used by individuals. Nimz et al. (1988) noted that these variations could be related to anthropometry. Belokovsky and Ivanchenko (1975) progressed this in an analysis of the amplitude and force of leg actions employed in breaststroke, establishing a range of hip and knee angles used by males and females throughout the leg kick. This group of authors also established that the average angles used by swimmers has changed since the 1950's and 1960's with increases in the angle of flexion of the hip joint during the leg recovery phase (95-120° to 130-140°) and the angle of flexion of the knee joint during the preparatory leg phase has increased (from 28-32° to 35-40°) (Belokovsky and Ivanchenko, 1975). The various hip and knee angles were then compared to the forces created by the swimmers and these authors established that the best combination of joint angles was 140° during hip flexion and 5060° knee flexion (Belokovsky and Ivanchenko, 1975). This was associated with an ideal SF of 65 cycles/min or 1.08 cycles per second and any changes outside this resulted in substantial changes in technique and a decrease in swimming speed. However, the substantial changes were not detailed. Belokovsky and Ivanchenko (1975) also attempted to ascertain whether training these specific angles altered the swimmer's performance and resulted in significant improvements in swimming time over a competitive distance. Unfortunately, there was no detail on how the swimmers learnt these new angles or how long the training duration was. There is a lack of literature on the kicking actions of the feet or whether this differs during training. This is important in developing the appropriate kicking technique to attempt to maintain during training to maximise training outcomes.

2.1.4. Summary

A range of kinetic technical variables, linked to mechanical efficiency and energy expenditure, are considered important for placing limbs in effective positioning, keeping the body in lateral alignment and decreasing resistance (Hay, 1993, Maglischo, 2003, Seifert and Chollet, 2005, Tella et al., 2008). Therefore, the specific kinematic technical actions a swimmer uses have vital consequences and importance for achieving maximal swimming velocity and optimal performance by influencing the factors that govern swimming speed. Discrete phases within the stroke cycle are often identified in the literature; however, there is a lack of literature pertaining to the specific kinematic parameters at particular events during a stroke cycle. The research to date has predominantly focused on race or competition outcomes, with only the study by Belokovsky and Ivanchenko (1975) relating their work to training environments. They have also used methods which cannot be used on a regular basis by coaches and are not conducive to use in a training environment. As these skills are learnt and refined through hours of practice and repetitive actions during training, it is imperative to apply this work into a training context to understand how coaches educate their swimmers on these factors, monitor their performance of them during training, and correct any errors to maximise their effectiveness in a race.

2.2. BIOMECHANICAL FATIGUE IN SWIMMING

Training sessions are designed by coaches to develop and hone swimmers' technical performance through repetitive practice, as well as challenge and push an athlete to achieve their potential in competition. However, during phases of high-intensity training, a number of factors may affect the manner in which technical actions are performed. One such stressor is fatigue. Despite the importance of technique in overall swimming performance, a limited amount of research into the effects of fatigue on technique has been conducted in a training scenario and so it is an important issue that remains to be investigated.

Although the focus of the current research is the biomechanical effects which occur with fatigue, these effects are thought to be related to changes in other underlying bodily processes. To inform and interpret the results of the present research, and compare them to other literature, some understanding of these processes is needed. Therefore, the mechanisms of fatigue are briefly reviewed.

2.2.1. The Mechanisms of Fatigue

Fatigue has been a major research topic since the work of Mosso and Hill in the late 1800s (Phillips, 2015), yet despite advances in technology and knowledge, the how and why of fatigue is still continually under debate and often individuals are unable to state with certainty why individuals become fatigued (Enoka and Duchateau, 2008). This appears to be associated with the numerous and varying definitions associated with fatigue, as mentioned in Chapter 1, which has clouded scientific enquiry into this topic (Marino et al., 2009). As a result, many researchers and sport scientists have undertaken varying individual approaches in trying to explain its underlying cause (Noakes, 2000) and in so doing, highlighted a range of potential mechanisms of fatigue which span many areas of sport science, including: physiological, psychological, biochemical, nutritional and biomechanical mechanisms (Abbiss and Laursen, 2005, Noakes and Gibson, 2004).

In an attempt to address this singular approach, a number of authors have attempted to devise a complex systems model of fatigue which combines many models to explain the mechanisms of fatigue (Abbiss and Laursen, 2005, Gibson and Noakes, 2004, Lambert et al., 2005). One such model, proposed by Abbiss and Laursen (2005), comprises nine submodels suggested to explain the mechanisms of fatigue which occurs during exercise (See Figure 2.2). These include: cardiovascular/anaerobic model, energy supply/energy depletion model, neuromuscular fatigue model, muscle trauma model, biomechanical model, thermoregulatory model, psychological/ motivational model and the central governor model. Each model will now be briefly discussed.

Firstly, the cardiovascular/anaerobic model, initially recognised by Hill and his colleagues in the early 1900's, suggests that exercise performance is limited by both the ability of the heart to supply sufficient oxygenated blood to the working muscles and the ability of the cardio-vascular system to remove accumulated metabolites (Abbiss and Laursen, 2005). The exact effect of accumulated metabolites and its role during exercise is also a factor of this theory which is keenly debated (Nielsen et al., 2001, Noakes and Gibson, 2004). This relates back to the belief in the 1900's that lactic acid produced during exercise and accumulation of lactic acid causes fatigue (Noakes and Gibson, 2004). This model predicts that training will increase the body's maximum capacity to consume oxygen resulting in an increased cardiac output and capacity of the muscles to consume oxygen and is argued to delay the onset of skeletal muscle anaerobiosis (Abbiss and Laursen, 2005). The key criticisms of this theory are that: the first organ to be affected by the oxygen deficit would be the heart, not the skeletal muscles (Noakes, 2000); a plateau in cardiac output must be developed before skeletal muscle anaerobiosis can develop, requiring a myocardial ischaemia which have never been shown to develop in

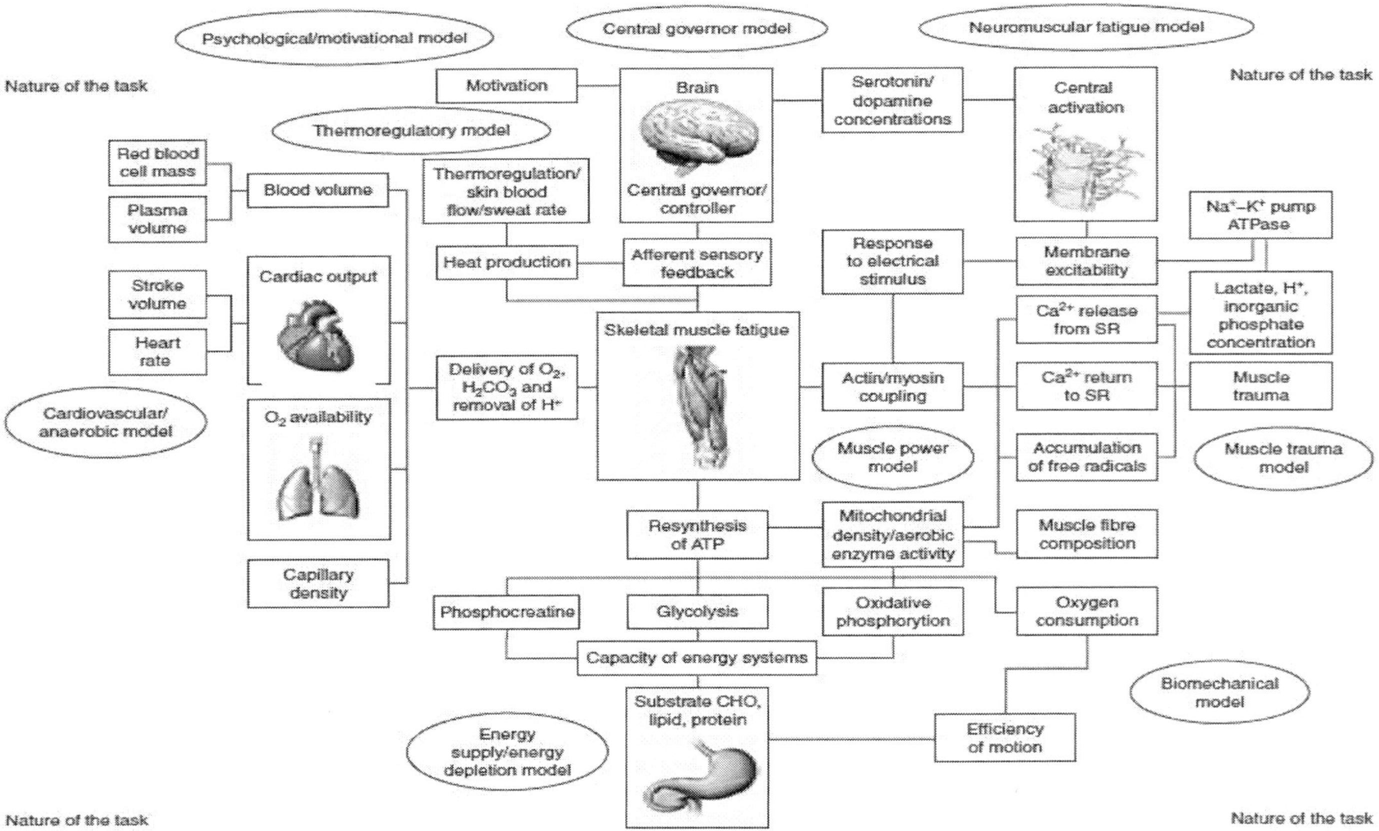

Figure 2.2: A diagram of the complex systems model proposed by Abbiss and Laursen (2005). ATP = adenosine triphosphate, ATPase = adenosine triphosphates; CHO = carbohydrate; SR = sarcoplasmic reticulum.

healthy humans (Phillips, 2015); changes in metabolite concentrations are difficult to measure and differentiate with fatigue (Noakes, 2000).

The energy supply/energy depletion model is related to the initial model and predicts that exercise will terminate when adenosine triphosphate (ATP) depletion occurs and is determined by the capacity to produce energy by different pathways (Abbiss and Laursen, 2005, Noakes, 2000). ATP is the energy source for the contraction of muscles and many other physiological mechanisms within the body. ATP depletion can occur when there is an inadequate supply of ATP within the working muscles, an inadequate supply to the working muscles using the various metabolic pathways (phosphocreatine system, anaerobic glycolysis and aerobic glycolysis) or as a result of the depletion of endogenous substrates (Abbiss and Laursen, 2005). It is suggested that this may arise either from a depletion of substrates (muscle and liver glycogen and glucose) or restrictions in the oxidative or glycolytic energy supply pathways (Noakes, 2000). This relates to the previous model hypothesis which holds that when the rate of ATP production by oxidative sources becomes inadequate, high rates of anaerobic glycolytic ATP production creates metabolites which interfere with energy production and crossbridge cycling causing fatigue and failure of muscle contraction due to peripheral inhibition. A criticism of this theory is that it predicts exercise must terminate when muscle ATP depletion occurs yet muscle ATP concentrations have been shown not to drop below 60% of their resting values (Green, 1997, Westerblad et al., 2010). Exercise can often continue to the point at which it compromises liver and glycogen supply despite sufficient oxygen (Abbiss and Laursen, 2005).

The neuromuscular fatigue model is a theory which postulates that the functions involved in muscle excitation, recruitment and contraction are what limit exercise performance due to a reduction in the force or power output of a muscle, despite increases in perception of effort (Cairns et al., 2005). This theory is based on the notion that in the chain of command from the motor centres in the brain to the actin-myosin cross-bridges in the muscle fibres, failure or impairment occurs (Gabriel et al., 2001a). Theories to explain this include the central activation failure theory, neuromuscular propagation failure theory and muscle power/peripheral failure theory (Abbiss and Laursen, 2005). The central activation failure theory involves a reduction in neural drive; neuromuscular propagation failure theory holds that fatigue results in reduced responsiveness of muscle to electrical stimulus; the muscle power/peripheral failure theory states fatigue occurs within the muscle and the excitation-contraction coupling mechanism (Abbiss and Laursen, 2005). Thus changes in the central neurotransmitters induce fatigue simply as a natural consequence of prolonged exercise and changes in relative balance of different neurotransmitters in the brain (Gabriel et al., 2001b). A criticism of this theory is that it is difficult to show the individual steps in the muscle activation sequence and therefore at present it is not possible to know which steps in this process are affected by fatigue (Abbiss and Laursen, 2005). An alternative side of this argument is that a reduced central activation of muscles is a necessary protective mechanism of the body and is related to the central governor theory described below (Noakes, 2000).

The muscle trauma model proposed that prolonged exercise may result in physical damage to the muscles, resulting in intra-muscular chemical changes which may in turn disrupt the muscle and lead to reduced neuromuscular activation and force production by the muscle (Nosaka et al., 2003, Enoka and Duchateau, 2008). The precise mechanism responsible for a decrease in performance of pain-induced muscles is unknown and research to date has focused predominantly upon physiological responses to eccentric exercise (Abbiss and Laursen, 2005). The significant disruption to the muscle may cause a reduction in neuromuscular activation and reduced force production of muscle which in turn can influence the capacity to perform sporting tasks (Abbiss and Laursen, 2005).

The previous models have focused on mechanisms of fatigue from a single sport science discipline. Conversely, the following five models begin to develop theories which comprise of multiple mechanisms from multiple disciplines of sport science. The biomechanical model suggests that fatigue leads to deterioration in movement coordination resulting in increased metabolic energy requirements for the same intensity of exercise (Abbiss and Laursen, 2005). Thus, greater demands are placed upon other physiological mechanisms leading to the changes suggested by the other models (Noakes, 2000). This model assumes that fatigue is governed by the efficiency of movement patterns and an improvement in movement efficiency would lead to a reduced VO_2 to handle a given workload, delaying the accumulation of metabolites and attenuated rise in core body temperature (Abbiss and Laursen, 2005). It suggests that the more economical the athlete, the longer they will be able to perform and the less demand on the other mechanisms responsible for fatigue before reaching failure.

The thermoregulatory model suggests that fatigue is associated with thermoregulatory responses that accompany prolonged exercise. The elevation in core temperature which accompanies prolonged exercise is thought to increase the physiological demands placed upon other systems and again, lead to the changes suggested by the other models (Abbiss and Laursen, 2005). The negative effects of environmental heat and hyperthermia on exercise performance have been well documented (Casa et al., 2005, Maughan, 2012, Sawka et al., 2007). However, the effects of environmental heat, dehydration and hyperthermia, and these mechanisms as causes of fatigue, are inconsistent in the literature, with both peripheral and central factors associated with hyperthermiainduced fatigue (Nybo, 2008).

The psychological/motivational model suggests that a psychological mechanism mediated by physiological inputs results in fatigue. This theory is supported due to the lack of evidence showing a single physiological variable is responsible for the adaptations resulting from fatigue (Abbiss and Laursen, 2005). Marcora et al. (2009) provided experimental evidence that mental fatigue limits exercise tolerance in humans through higher perceptions of effort rather than physiological mechanisms. This fits with 'Brehms theory of motivation' that subjects decide to withdraw effort when a task is perceived to be either too difficult or effort demands exceed the upper limit of what people are willing to do, as well as that performance can be sustained or improved from verbal encouragement. As the precise mechanism that affects the brain's response to afferent feedback is unclear, it is therefore suggested that several physiological mechanisms lead to psychological changes which determine the unconscious perception of fatigue, leading to reductions in the intensity of activity that is maintained and, potentially, the attainment of a point of exhaustion (Noakes, 2000). This provides evidence of a physical-cognitive effort and the potential involvement of multiple mechanisms from several disciplines of sport science.

Finally, the central governor model proposes that the origin of fatigue is located within the central nervous system (CNS) with a loss of muscle force occurring through processes proximal to the neuromuscular junction, specifically within the brain, spinal nerves and motor neurons (Abbiss and Laursen, 2005). A common approach to analyse this theory was the comparison of maximal voluntary contractions created by electrical stimulation (Enoka and Stuart, 1992, Vøllestad, 1997). This suggests that peripheral inputs from the active muscles, chemo and thermoreceptors, and other areas are integrated by a central mechanism and that exercise intensity is mediated to optimise exercise performance, as noted by Ulmer (1996), and as a safety mechanism which preserves metabolism, cellular integrity and muscle capabilities, as expanded upon by Noakes et al. (2001). However, some of these methods are questionable, with issues including the necessity for a well-motivated individual to maintain a maximal CNS drive and the difficulties of assessing someone's motivation or whether someone has full motivation to perform (Davis and Bailey, 1997). These issues strongly emphasise the issues associated with measuring CNS fatigue, but objectivity in measurements are a vital aspect of assessing an individual's subjective willingness to continue exercising (Phillips, 2015). Noakes et al. (2001) have re-introduced this concept of mind-body interactions through this theory.

It is apparent that the models of fatigue can be identified in two main areas, highlighting one of the biggest debates in research in fatigue which has existed for over a century; whether the aetiology of fatigue is central (relative to the motor neurone) or peripheral (relative to motor fibres) (Kay et al., 2001). Each area is supported by rigorous studies which have proven that both can occur as well as limitations in their applicability to sporting performance. However, the CNS theory did not become popular until the past few decades (Phillips, 2015). This may have been due to the popularity of the peripheral theory and the limitations of the capacity to measure central fatigue and test the underlying models properly (Phillips, 2015).

The physiological, biochemical, biomechanical and cognitive models used to explain fatigue are diverse. Taken individually each model has potential flaws and failings in terms of application to real life and they are too simplistic to adequately explain the complexity of fatigue during exercise. As Noakes (2000; p. 124) stated "*it is highly improbable that the factors explaining human exercise performance under all conditions are restricted to one physiological system*". The complex systems model attempts to account for these flaws and addresses these issues by considering the multi-factorial nature of fatigue and the interaction of the numerous bodily systems and mechanisms previously described. Due to its multi-factorial approach and the apparent multifactorial nature of fatigue, the present thesis will follow the complex systems model approach. However, it is important to note that at present the aetiology of fatigue remains an important yet controversial area of research (Fitts, 2008).

The present study does not intend to evaluate these theoretical models nor ascertain which are pertinent to the present research due to the range of literature and lack of knowledge of precise factors that determine fatigue and limit performance (Noakes, 2000). However, the mechanisms which contribute to fatigue are influenced highly by the type and intensity of the exercise, the actions or activity being performed, the individual and physical environment and thus any effects of fatigue are likely to vary according to the specifics of the exercise task (Knicker et al., 2011). Thus, even subtle changes in a task can be associated with marked differences in the time to failure (Hunter et al., 2004) and to better understand the influence of fatigue on swimming technique, understanding the progressive effects of fatigue on technical performance is the focus of the present study.

2.2.2. The Effects of Fatigue on the Kinematics of Movement

One aspect which is agreed on in the literature is that the site of failure and subsequent effects of fatigue are dependent on the task and the dominant mechanism is also specific to the processes stressed during the activity (Enoka and Duchateau, 2008, Knicker et al., 2011). The contribution varies due to a dependence on task factors, including: the task duration or intensity; and individual factors including: differences in aspects such as age, gender, fitness and diet (Enoka and Duchateau, 2008, Fitts, 2008, Kay et al., 2001). Thus fatigue effects are individual and sport specific and should be dealt with accordingly. Effects of fatigue commonly reported in the literature can be wide ranging and include physiological, biochemical, psychological and biomechanical effects (Ament and Verkerke, 2009, Enoka and Duchateau, 2008, Knicker et al., 2011). However, as the focus of this thesis is upon kinematic parameters only literature pertaining to effects of fatigue on kinematic variables associated with technique will be reviewed here.

2.2.2.1. Effects of fatigue on technique in other sports

Fatigue has been a topic of research and important part of athletic performance for many years. Individuals involved in sports performance have been aware that athletes' technique can change as they become fatigued. As these changes in technique appear during on-going exercise and the skill can continually be completed (due to the varied ways the body's muscular system can move), this does not indicate only a reduced motor skill capacity (Knicker et al., 2011). According to Knicker et al. (2011; p. 313), the same skill outcome "*can be achieved with different patterns of movement coordination, i.e. technique deviation rather than deterioration*". Royal et al. (2006) emphasised this by showing that technical water-polo skills decreased yet shot speed and accuracy was unchanged during intense training. Technique deviation is the change in technique from the 'norm' without an impact on the final outcome of the action (Knicker et al., 2011). This is theorised to occur as even when fatigue occurs in one muscle, the synergist muscles work together to sustain the work intensity, prevent loss of overall power and continue performance of the action (Knicker et al., 2011). This has been shown in elite handball players who were able to maintain the hand velocity yet showed changes in coordination during repetitive throwing (Forestier and Nougier, 1998). Technique deterioration on the other hand results in greater changes of motor skill execution and diminished outcomes of the action (Knicker et al., 2011) and has been shown in a number of actions, including throwing and kicking (Kellis et al., 2006). Despite these findings in basic movement actions, far less research has been conducted in sporting performance. Of the work that does exist, it is sporadically spread throughout a range of sports or activities, including:

- Running: There is a dearth of literature on running and fatigue and changes in technique including: kinematic leg actions; decreases in step length; increased cadence and an altered ground reaction force (GRF); changes in bone strain patterns; and alterations in strain location (Christina et al., 2001, Derrick et al., 2002, Gerlach et al., 2005). A range of test sets and running distances have been used in the literature to assess such changes, including running a set distance or time to failure sets (Christina et al., 2001, Morin et al., 2011) and completing distances during such sets ranging from 440yds to 31 miles. Despite this range of literature, no studies have been conducted to analyse the running technique during training scenarios.
- Rowing: Pollock et al. (2012) analysed the kinematic changes during 2000m but provided very little information throughout the test set nor analysed individual differences.
- Cycling: the limited number of cycling studies have focused on lower limb angles of the main leg joints (hip, knee and ankle) during a leg cycle, predominantly at a range of various percentages of the cyclist's aerobic capacity (Amoroso et al., 1993, Delextrat et al., 2005, Dixon et al., 2013, Sarre et al., 2005). One study in cycling has been conducted to investigate the cumulative effects of training stress and recovery on performance changes by monitoring and controlling the training of eight endurance cyclists over two weeks (Halson et al., 2002). This study was one of the first attempts to systematically induce a state of overreaching while monitoring stress and performance in a highly controlled environment. Unfortunately this was only assessed using a single time trial event at the end of the phase and no changes were analysed during the two weeks of training.
- Rugby: The focus of the limited research in rugby has been on specific match actions such as tackling (Gabbett, 2008) or kicking (Coventry et al., 2015). Coventry's analysis of punt kick

kinematics noted that an increase in the velocity and range of motion of the hip and pelvis was an adaptation (or technique deviation) in the performance to cope with the fatigue in the short-term (Coventry et al., 2015). On the other hand, Gabbett (2008) noted that a player's tackling technique severely reduced and they were far less effective at performing a safe, successful tackle (or technique deterioration had occurred) after fatiguing sprint sets.

- Water-polo: Oliveira et al. (2015) analysed changes in the egg beater kick during a fatigue set and noted that the speed of feet, hip abduction and flexion decreased with fatigue while hip internal rotation, and ankle inversion increased.
- Other sports, including tennis (Hornery et al., 2007), basketball (Nezhad et al., 2015) and soccer (Chan et al., 2014) have also shown similar changes in technical parameters due to fatigue during game- or match-like scenarios but no research has been conducted in training situations.

Despite the dearth of literature, many sports remain to be analysed in terms of the effects of fatigue on biomechanical technical actions. Given the dissimilarities between swimming and other sport activities in terms of the environment and technical actions, the conclusions that can be drawn are limited. The main point to take from this literature is that fatigue can have a range of effects on technical and biomechanical factors in a range of skills, activities and sports, both on specific technical actions as well as entire body movements. Although informative, these studies utilise a range of participants, from non-trained participants to elite athletes, and therefore do not appear to take into consideration the individual differences in technical actions as well as effects of fatigue. Further research is required to analyse these in more detail. The review of such literature is therefore brief and a more comprehensive review of the limited work relating to the effects of fatigue on swimming technique follows.

2.2.2.2. Effects of fatigue on technique in swimming

Although fatigue has been mainly examined using a wide range of physiological, kinematic, biomechanical parameters, this review focuses on kinematic effects of fatigue only, including SL and SF.

The changes that occur with fatigue in swimming have been investigated since the early 1970's with investigations into competitive events being studied extensively (Craig and Pendergast, 1979, Craig et al., 1985, East, 1970, Pai et al., 1984, Sidney et al., 1999). These studies investigated a range of events, usually at elite national and international competitions, and showed that the decrease in swim times from 1976 trials to 1984 trials was due to increases in SL. In almost all events the male finalists had a greater SL and reduced SF in comparison to the women or slower counterparts. Research by Craig and Pendergast (1979) highlighted that women relied more on SF than men in all four competitive strokes. Similar changes in stroking parameters were also validated in frontcrawl in a much larger population of male swimmers (n=442) of differing skill levels in the 100m front-crawl (Chollet et al., 1997). Since that time SL and SF were shown to be able to discriminate between the performance levels of swimmers, with elite swimmers able to maintain SL and SF more effectively and have a smaller decrease in SL than novice swimmers during race performances. The smaller decrease in SL has been linked to a smaller decrease in power output in elite swimmers (Monteil et al., 1996) and other differences in stroking parameters and technique styles have also been related to anthropometric factors (Persyn et al., 1975, Soons et al., 2003).

These studies were the first to identify that as a swimmer tires during the race, there is a loss of SL related to a decreasing ability to develop the force necessary to overcome resistance to forward movement. During the 1990's, it was suggested that the decrease in SF could be related to decreases in force production or diminishing capacity to deliver power output (Toussaint and Beek, 1992). This was further investigated by Toussaint et al. (2006) who investigated the changes in speed, SL and SF with mechanical power output during a 2x100m front-crawl. The decline in SF noted during the swims was thought to reflect a reduced propulsive force and the decline in speed a decrease in the power producing capacity of the swimmer. Toussaint et al. (2006) suggested that swimmers attempt to maintain optimal power output by optimising the SL-SF combination to maintain the highest possible SL while swimming. Further research is needed to ascertain whether an optimal SL-SF combination can be achieved and whether this can be trained and monitored during training.

Another possible factor contributing to the decrease in speed with fatigue is related to the decrease in neural activation failure linked to physiological factors (Keskinen and Komi, 1993). As a consequence of the suggestion that the stroking parameters in swimming were influenced by physiological factors, there has been a surge of research into the association between stroking parameters, the intensity of the swimming workload, and changes in metabolic variables. Wakayoshi et al. (1996) used a swimming economy test of 4-6 x 400m swims followed by two additional 100m swims at a swim speed corresponding to each athlete's maximum oxygen uptake (VO_2max) to assess changes in swimming mechanics (stroking parameters) at different workloads. They noted that for the submaximal efforts (aerobic workload), SL and SF remained unchanged; however, during swim speeds above the onset of blood lactate accumulation, SL decreased and swimmers increased SF to maintain constant velocity. Similar findings were established by Laffite et al. (2004) during analysis of a broken down 400m swim and Dekerle et al. (2004) in a straight 400m maximum effort swim. Conversely the changes in the stroking parameters during the maximum effort 400m swim were noted in only five of the total eleven swimmers and highlighted that the "*evolution of the stroking parameters, when the speed is increased, depends on the task constraints*" (Dekerle et al., 2004; p. 56). These differences identified the individual differences among swimmers in their reactions to fatigue, including the effects of different workload intensities on stroking parameters. As a result Dekerle et al. (2004) suggested that swimmers should resist the reduction in SL during training and advised coaches to perform simultaneous measurements of technique variables and physiological parameters under physiological stress and to control training stress. Unfortunately, within these studies a range of different distances were analysed, with a predominant focus on race events or 400m swims. The authors all agree that at higher intensities of swimming, stroking parameters can be affected. However, the swimming intensity is often defined and measured in ways that differ among studies, including different lactic acid thresholds or VO_2 values (Dekerle et al., 2004, Keskinen and Komi, 1993, Laffite et al., 2004). Although a range of intensities were used, none of these resembled those experienced in training, during which swimmers push themselves to their physical limits to prepare for competitive practice.

A range of studies have shown a decline in swimming speed during competitions and linked these to changes in SL, given that SF remained stable during the races. For over three decades the focus of research on fatigue in swimming was on the effect of fatigue on stroking parameters. In 1996, Monteil verified that the changes in force distributions and propelling efficiency occurred at different phases of the stroke cycle as a result of fatigue and suggested that these changes reflected swimmers modifying their coordination in order to maintain the most effective combination of stroking parameters to maximise their swimming speed. Following this study, and the development of the IdC method by Chollet et al. (2000), researchers began to focus on the effects of fatigue on stroking parameters and their relationship to stroke coordination. Although Chollet and colleagues have been at the forefront of the understanding of swimming coordination and influential factors, Alberty and colleagues have completed a large range of research into the effects of fatigue on coordination using the IdC method. Alberty and Colleagues have assessed the effects of coordination during freestyle 400m time to exhaustion test sets during free or controlled swimming speeds (Alberty et al., 2008, Alberty et al., 2009, Alberty et al., 2011). These authors attempted to discover the strategies adopted by swimmers to manage their propulsive impulses as fatigue develops (by manipulating stroking parameters and workload intensities) in an effort to provide a better understanding of the observed changes in stroke technique (Alberty et al., 2009).

The research by Alberty and colleagues identified similar changes in stroking parameters as the previous studies on stroking parameters during competition events (Craig and Pendergast, 1979, Pai et al., 1984) in that at aerobic paces swimmers are able to maintain their SL-SF combination but at anaerobic paces this combination must change to maintain an imposed speed. This change is observed as an increase in SF and a decrease in SL when swimmers are free to choose their own SL-SF combination (Alberty et al., 2008). In addition, it was noted that the duration of the propulsive phase in the stroke increased and was thought to be due to an increase in SF and a consequential increase in the total stroke cycle duration (Alberty et al., 2009). This was linked to some of the mechanisms of fatigue previously discussed, including exhaustion in the working muscle groups (Abbiss and Laursen, 2005). However, when a stroking parameter was fixed or speed was constant it was identified that tests to exhaustion were longer and variables were consistent suggesting a stabilisation of stroke technique. This supported the hypothesis that '*at a constant speed a controlled SF would induce stabilization of the temporal structure of the stroke cycle*' (Alberty et al., 2008; p. 1192). Without manipulating the stroking parameters the time to exhaustion was shorter and related to the intensity imposed and associated with a slower SF. Craig et al. (1985) and Alberty et al. (2011) suggested this may be due to an inability to minimise the resistive forces and that, due to fatigue, swimmers may pay less attention to their body alignment and therefore swim with a less streamlined position. At present, no research has investigated this.

From the research on fatigue, coordination and stroking parameters in swimming, some suggestions have been offered to coaches in terms of training formats in that if coaches want to stabilise a swimmer's technique and coordination of propulsive actions they should control the swimming speed and SF (Alberty et al., 2008) or to also improve propulsion, coaches could constrain swimmers to increase their SL at a given velocity

Alberty et al., 2005). As there have been no studies assessing these methods during training (whether acute or long-term) it would be advisable to use these methods carefully in training (Alberty et al., 2005). In addition, the previously described research has some limitations in terms of application in that they focus on front-crawl swimming and none have been conducted in the other strokes. Also, only recently have studies included both male and female swimmers. Although individual differences have been noted in terms of intra-cycle speed and SL-SF combinations (Alberty et al., 2011), no research has been specifically conducted to analyse these components on an individual basis.

Due to the significant role training plays in learning the technical skills required for performance, an understanding of the effects on stroking and kinematic parameters is essential for athlete development and the maximisation of the training time. Due to the initial focus on 400m tests to race-like distances in research on fatigue in swimming, only a small number of studies have investigated the effects of fatigue on stroking parameters and kinematic variables during training-like sets. These have included sets such as: a 7x200m set (Marinho et al., 2006); 4x50m swim sets (Aujouannet et al., 2006); 6x50m swim set (Deschodt et al., 1999); 10x50m and 100m sprint (Stirn et al., 2011); and an interval set of 5x400m swims at a 200m-400m pace (Ribeiro et al., 2010). However, these studies were all predominantly of front-crawl, used male participants only and did not analyse the individual differences in these parameters during the sets. Similar findings were identified in terms of changes in stroking parameters to those studies mentioned previously (Alberty et al., 2009, Craig and Pendergast, 1979, Pai et al., 1984, Toussaint et al., 2006) highlighting a reduction in the quality of the stroke, represented by changes in stroking parameters, associated with a lower capacity of force production to overcome water resistance (Ribeiro et al., 2010).

Alberty et al. (2011) suggested that knowledge on how swimmers adapt their technique during training as a result of fatigue would be of benefit to coaches to help guide their training programmes and feedback. However, Alberty et al. (2011) also noted that the technical variables and acute changes which occur remain unknown and are still unclear, as the predominant factor in research has been stroking parameters. Studies of the effect of fatigue on stroking parameters have revealed the importance of kinematic parameters yet very little research has been conducted to investigate these topics. Due to the importance of hand and arm actions during swimming, existing research into the technical variables have focused on hand and arm kinematics. Decreases in the hand speed and wrist displacement have been noted during 400m, 200m and 6x50m frontcrawl maximal effort swims (Deschodt et al., 1999, Figueiredo et al., 2013, Monteil, 1994). Martins-Silva et al. (1997) also noted this during the upsweep phase in 200m butterfly. Several studies have noted very little change in the trajectories of the hand during underwater phases and ascertained that the "*maintenance of the hand trajectory suggests a robust spatial pattern in these elite swimmers that is not easily changed even by the impairments imposed by fatigue*" (Aujouannet et al., 2006; p. 155) and changes were therefore related to an inability to maintain a high force output (Monteil, 1994). As neither the force output nor spatial patterns were directly measured, neither can be conclusively confirmed. In addition, Suito et al. (2008) identified that shoulder adduction and hand speed reduced statistically significantly during the pull phase over a 100m front-crawl and this was compensated by increases in internal rotation. One reason that a 'robust hand trajectory' could be maintained is that alterations in technical variables in other areas of the body during the stroke cycle compensate to maintain an optimal swimming speed (Suito et al., 2008, Suito et al., 2007). Due to the individual differences it cannot be assumed that female swimmers would have the same changes as male swimmers nor that males of different levels or in different strokes would have similar changes. Although these data are informative, the methods utilised are complex and time-consuming with large mathematical contributions or the use of expensive camera equipment and 3-D analysis software (Eltoukhy et al., 2012, Kirmizibayrak et al., 2011). Unfortunately, it is unknown whether these changes can be observed visually or monitored using coaching methods during training and therefore, whether this information can be used by coaches to monitor technique during training.

As a result of the focus of the literature on front-crawl, studies regarding the effect of fatigue during breaststroke swimming, either in race or competitive scenarios, are scarce or have focused only on changes in stroking parameters during race-like events. According to the literature in 200m events, the swimmers SL-SF combination is very individual with some swimmers possessing a high SF and a low SL while others swam with a high SL and low SF (Chatard et al., 2003, Conceição et al., 2014, Leblanc et al., 2010). No reasoning was provided for these individual differences in terms of kinematic differences in the technique style. The limited studies on breaststroke swimming technique arrived at similar conclusions regarding the effect of fatigue on the stroking parameters during 200m events in that speed and SL decrease, whereas SF decreases then increases (Craig and Pendergast,

1979, Takagi et al., 2004, Thompson et al., 2000, Thompson et al., 2004). Due to the nature of the breaststroke technique and individual differences, further research in this technique is needed to understand the effects of fatigue on breaststroke technique.

2.2.3. Summary

The literature on fatigue in swimming has an overwhelming predominance on stroking parameters, coordination or, at most, some upper limb kinematic technical actions. These have focused mainly on male participants, front-crawl technique and race-like scenarios. One main outcome from these studies was the link to the importance of reducing resistive forces and maximising propulsive forces which ultimately is related to the technical actions a swimmer performs. It is during training that these technical actions are learned and honed to enable the swimmer to maximise their performance. Unfortunately, literature pertaining to the specific and individual kinematic changes which occur during training as a result of fatigue or high-intensity training are very scarce. The paucity of research in training situations therefore leaves many questions unanswered. Without this information, it is difficult to ascertain whether fatigue related changes in technique occur progressively during training exercise and whether these can be observed and monitored by coaches. Therefore, the work that exists possesses limitations and methodological flaws and further investigations in this area are warranted.

2.3. THE ANALYSIS OF TECHNIQUE IN SWIMMING

According to Lees (2002; p. 814), analysis of technique has many definitions but is frequently referred to as '*a prerequisite to the process of improved performance*' and is used to '*improve technique*'. Although practiced for many years it was first developed into an organised process based on scientific principles in the 1950's (Lees, 2002). The term is also known as: 'the analysis of sports techniques' (Bunn, 1972); 'biomechanical analysis of technique or movements' (Adrian and Cooper, 1995, Bartlett, 2002); and 'the analysis of sports skills' (Carr and Carr, 1997). Although these terms vary, the purpose and processes involved are relatively the same (Lees, 2002). Several issues have been identified in the literature regarding the processes of technique analysis and include: the range of definitions and therefore analysis methods of technique (Bartlett, 2002, Bober, 1981, Carr and Carr, 1997, Elliott, 1999); the definitions described above often do not establish the scope of the technique or different individual styles which may exist between athletes (Lees, 2002); often the analysis does not establish or quantify whether the technique was effective or in other words whether the athlete completed the skill correctly using the most efficient actions (Lees, 2002).

Despite these issues, technique analysis is a well-used method in sports performance. It evolved based on the application of mechanical (or biomechanical) principles of movement (Lees, 2002) and their application to performance skills and has developed into a range of approaches to analyse technique. These approaches have resulted in two extremes of technique analysis methods, from the qualitative perspective of making observations regarding a movement to the quantitative perspective of using carefully designed experiments or computer models to simulate interventions and analyse technique (Lees, 2002). This thesis focuses on the qualitative and quantitative approaches of technique analysis only, as it is focusing on the observation or recording of real movements, not the simulation of body models (or predictive analysis). These two approaches are now discussed.

2.3.1. Qualitative Approach

Qualitative analysis is one of the most important activities of coaches in sport performance. Knudson and Morrison (2002; p. 4) defined it as the "*systematic observation and introspective judgement of the quality of human movement for the purpose of providing the most appropriate intervention to improve performance*". It was used initially when there were few observational and analytical aids available and, although it is based on scientific principles, uses subjective observation to identify and correct faults in skill execution (Knudson, 2007, Lees, 2002). Qualitative analysis is often used in the coaching process in training and competitions as a means of obtaining feedback to help plan programmes and prepare the athlete to compete (Mets et al., 2003). There are several well-documented systematic approaches to the qualitative analysis process (Lees, 2002).

To aid the process of qualitative analysis of technique, a large range of models have been proposed. These models have been classified as comprehensive models and provide a summary of the components necessary for qualitative analysis (Knudson and Morrison, 2002). Comprehensive models can be further broken down into pedagogical models (Arend and Higgins, 1976) and biomechanical models (Kreighbaum and Barthels, 1996, Lees, 1999). These models are intended to make the process of fault identification and performance enhancement more manageable and help individuals to analyse sport movements. However, the pedagogical model developed by Arend and Higgins (1976) was limited in that it was predominantly descriptive, could not practically measure the kinematic or kinetic aspects of the action, and assumes that the observer has a detailed knowledge of the technique and biomechanics. Although the number of phases and names of each phase can deviate between models, Lees (2002) highlighted three main steps to the qualitative analysis approach common among comprehensive models: observation, evaluation and intervention.

2.3.1.1. Qualitative analysis: Observation

Technique observation is a common tool for gathering information about human behaviour and is often used in the coaching process during training and competitions (Mets et al., 2003). According to Mets et al. (2003) visual processes and perception are some of the main methods available to coaches when analysing performance in training and it is assumed that this requires many skills from the observer (Knudson and Morrison, 2002). While different senses provide unique information about a performance, vision is the most sensitive to spatial changes in the position of the human body. According to Moreno et al. (2006; p. 861), to be effective this requires "*the observer to focus attention only on the most relevant or crucial sources of information*". In order to achieve this, observation is often separated into three components: Phase analysis, temporal analysis and critical features (Lees, 2002).

- Phase analysis involves breaking a movement into a minimum of three relevant phases, such as preparation, action and follow through, so that attention can be focused on the performance of each part and is descriptive in nature (Kreighbaum and Barthels, 1996). These phases can also be broken down further into 'sub-phases' (Lees, 2002). It requires a basic knowledge of the skill being analysed and coaches should integrate experiential and sport-specific knowledge with the relevant biomechanical principles (Knudson, 2007).
- Temporal analysis was defined by Lees (2002; p. 816) as an '*attempt to specify the timing of a movement and builds into sequences of movement established through phase analysis*'. It involves aspects such as timing and rhythm yet despite its use by some authors is not often referred to in qualitative analysis (Adrian and Cooper, 1995, Lees, 2002).
- Critical features are any observable aspect of a movement which are deemed vital to the outcome of the performance or

skill (McPherson, 1990). This term was introduced by Arend and Higgins (1976; p. 45), who defined it as "*parts or phases of a movement which can be least modified to achieve a goal*". Critical features often consist of descriptions about the position, attention and movement of a skill which are described using the skill characteristics or biomechanical principles of movement (Lees, 2002).

These three components are thought to be starting points to aid practitioners in the qualitative process. However, the actual use of each component by practitioners appears to relate to the simplicity and ease of use of each method with phase analysis being very popular due to its limited need for preparation whereas critical features, which often require background knowledge of the skill, appear to be less well used (Bartlett, 1997, Knudson and Morrison, 2002, Lees, 2002).

A number of observation models have been developed to overcome the limitations of naked eye observation and to facilitate the identification of critical features during movement in sport. Some models break down the skill into phases and others allow the analyst to get an overall feel about the quality of movement (Knudson and Morrison, 2002). To translate into observable features, biomechanical observational models have often been developed. These are used to guide the analyser in the observation of specific technical and performance variables, both those mechanically related to performance and observable (Knudson and Morrison, 2002). Coaches are encouraged to use all of their senses in a systematic observational strategy of several trials of skill before moving onto the evaluation and the diagnosis phase of qualitative analysis (Knudson, 2007).

2.3.1.2. Qualitative analysis: Evaluation

Evaluation is the second step identified in the qualitative approach. It refers to the way in which subjective judgements are made about performance and how faults are diagnosed. It also has three main approaches: templates, principles of movement and systematic models (Lees, 2002).

- Templates are an idea or model which characterises the 'ideal' way in which a movement or skill is performed (Lees, 2002). It can be in a range of formats, including written or diagrammatic (Lees, 2002). In real-world settings this method is often achieved by the use of an expert athlete performing a demonstration (Lees, 2002). By watching a performance, the practitioner can identify errors or faults in the technique in comparison to the 'template' (Lees, 2002). However, this does not take into account individual differences or encourage critical thought and assumes that success equals high technical skill (Knudson, 2007).

- Principles of movement is the earliest and most widely used approach to qualitatively evaluate technique and is the application of biomechanical principles to technical performances (Lees, 2002). Many authors have attempted to identify principles of movement and a range of principles exist from which coaches can use (Bunn, 1972, Hochmuth, 1984, Hudson, 1995).
- Systematic models are widely used in biomechanics and have been described by a number of authors in qualitative analysis to evaluate important characteristics of skills, with the model by Hay and Reid (1982) held in very high regards (Lees, 2002). These models are known under a wide range of terms. This approach is based on a hierarchy of factors on which the results or outcomes of the performance depend (Lees, 2002). The main rule is that each of the factors in the model should be determined by those factors that appear immediately below it, either by addition or mechanical relationships (Hay and Reid, 1982). Consequently, the model can be used to identify factors relevant to performance but, unfortunately, not aspects of technique relevant to these factors (Lees, 2002). It is a complementary method to those previously described and requires the use of the previous methods.

Throughout this process, coaches identify strengths and weaknesses of performance and analysts tend to compare observed movement to expected or mental images of a movement or to relevant biomechanical variables or principles. Hay and Reid (1982) recommended the use of the deterministic models of movements as the best approach for qualitative analysis but regardless of the method, the coach makes a judgement about the quality of performance (Lees, 2002). This requires a thorough knowledge of the skill being analysed, the basic steps involved in observing and identifying the technical faults, prioritising the faults that need correction and finally instructing the performer.

2.3.1.3. Qualitative analysis: Intervention

During the final aspect, the intervention, the coach attempts to fix the diagnoses identified in the previous section; however, this process has received little research interest (Lees, 2002). Research evaluating the effectiveness of intervention methods in correcting technical diagnoses is also scarce (Lees, 2002). This lack of knowledge prevents deeper understanding of the value of these aids and qualitative methods used by coaches and whether they are applied during training or competition scenarios (Knudson, 2007).

2.3.1.4. The limitations of qualitative analysis

Traditional coaching often involves subjective observations and conclusions based on the coach's perceptions, biases and own previous experiences. However, a number of studies have revealed that subjective observations are potentially both unreliable and inaccurate (as highlighted in Chapter 3). Human memory systems have limitations and it is almost impossible to remember accurately all the meaningful events that take place during an entire competition, let alone provide any detailed analysis. In addition, a number of factors can influence vision, including: the speed of the movement, colour differences, the action itself, and eye variables such as eye dominance or peripheral vision (Knudson and Morrison, 2002). This may explain why observations may not be reliable nor valid in qualitative analysis. One reason for the limitations of the current qualitative approaches coaches use to analyse technique could be a result of the limited research in qualitative technique analysis methods used by coaches since the late 90's. The same methods established between the 1970's and 1990's are still referred to in current coaching and qualitative analysis literature (Knudson, 2007, Lees, 2002). Further research into qualitative technique analysis methods and the approaches coaches use may develop and potentially improve these limitations. Since successful coaching depends, among other things, on the accuracy of observation and how well it is analysed, it is extremely important that the information collected during athletic performance is objective, unbiased, and as accurate as possible (Hughes and Franks, 2007). This is discussed further in Chapter 3.

2.3.1.5. Summary

The coaching process can be thought of as an on-going cycle of performance and practice and part of the coach's role is to observe and analyse the performance of athletes and provide feedback, which can be incorporated into planned practice that should, theoretically, lead to enhanced performance. Coaches usually observe the whole body movement of an athlete, factorise this movement into specific components, and describe the athlete's technique as a combination of multi-segment movement components (Federolf et al., 2014, Knudson and Morrison, 2002). These recommendations and models, although often useful practical advice, are based on subjective observation and interpretation of athletes' movements and may be incorrect or not the best solution for an individual athlete (Federolf et al., 2014). The processes of observation and evaluation described in the literature is lengthy and quite complex and this appears to contrast the simpler approach more evident in sports coaching, based on the template of a model performance (Lees, 2002). The use of these models requires an accurate technical understanding of the skill, and thus, an underlying knowledge of the biomechanical principles and how they apply to the skill (Hughes and Franks, 2007). Therefore, qualitative analysis requires a range of skills and underpinning biomechanical principles in its use. However, the limited research regarding coaches' biomechanical knowledge, what methods coaches actually use, and how successful they are in aiding the coaching process remains unknown.

2.3.2. Quantitative Approach

Historically, technique analysis has been developed from some form of model, as described in section 2.3.1, to quantify the key biomechanical variables related to performance (Lees, 2002) and then used to provide some form of feedback to the athletes. As equipment and scientific analysis methods and procedures have developed, they have become more widely available for practical use and it has become possible to measure those aspects of skills related to technique, often using kinematic and temporal variables (Lees, 2002). Although subjective estimation for quantifying variables has been used with some success (Cutting and Kozlowski, 1977, Douwes and Dul, 1991), the usual quantitative approach uses some form of instrumentation. There are a large range of instrumented data collection methods for quantifying skill (such as those described in Section 2.2) and usually include motion analysis, force analysis and electromyography (EMG) (Knudson, 2007). These methods have been mainly used to describe the descriptive goal of quantitative technique analysis by sport biomechanists (Lees, 2002).

Quantitative analysis provides a different challenge, as these methods enable measurement of small details, which reflects essential characteristics of technique and which then need to be processed (Lees, 2002). These small details may be analysed by either cross-sectional (compare different sports performers techniques at a particular competition) or longitudinal processes (compare several trials of the same individuals) (Bartlett, 2007). Three main approaches are specified in the literature for the selection of technique variables: the use of deterministic models, reference to previous research and articles, and statistical models (Lees, 2002).

Following selection of the key variables, this approach then involves recording and interpreting measureable variables to quantify and analyse human movement and technique in sport (Hughes and Bartlett, 2002, Federolf et al., 2014). The overridingfocus in applied sports biomechanics has been the accurate quantification of movement kinematics and kinetics and the development and advancement of measurement and analysis systems (Lees, 2002). As a result, this approach is predominantly used by sports biomechanists in technique and performance analysis in sport. It does currently have some limitations. By singling out specific variables rather than considering the whole movement of an athlete, a sport scientist may miss important information and may not even be able to determine the origin of a change in observed variables (Federolf et al., 2014, Lees, 2002). In addition, due to the use of instrumentation, quantitative technique analysis has been associated with being time-consuming, requiring considerable set up and equipment and this has often hindered its application in actual sporting settings (Kerwin and Irwin, 2008, Lees, 2002). In particular, quantitative analysis techniques are rarely used by coaches.

2.3.2.1. Summary

Quantitative methods are powerful and have much promise for technique analysis. Although applied for some time they have only begun to make significant statements on sport technique. This is partly due to the time required to develop suitable data collection tools that can genuinely be applied to the study of performance skills but also by the lack of conceptual base for application of these methods in training settings by coaches.

2.3.3. The Relationship between Coaches and Sport Scientists

Technique and movement patterns are the components that enable an athlete to perform and thus are at the heart of sports performance. In order to improve sports performance in coaching, a crucial aspect coaches are expected to know is how to develop performance skills and movements to a more optimal technique through training. As a result of this, coaches are presumed to have a strong technical knowledge of the skill and a mental model of the correct performance of it (Smith et al., 2015). As mentioned in Chapter 1, sports biomechanics is defined as '*application of mechanical laws to living structures and biological systems, specifically the loco-motor system of the human body*' (Hay, 1993 p. 2) and is known as the underlying science behind technique (Hay, 1993). As the purpose of technique analysis is to improve performance and consider the diagnosis, identification and intervention of technique faults, knowledge of sports biomechanics and the principles underlying technique performance is thought to play a vital role in coaching. (Federolf et al., 2014, Lees, 2002). Therefore, knowledge and understanding of sports biomechanics provides a means that may help coaches better understand technique by helping to: identify the most effective skill development pathways; reduce the risk of injury; and remove the trial by error nature of training by making it more effective and efficient (Kerwin and Irwin, 2008).

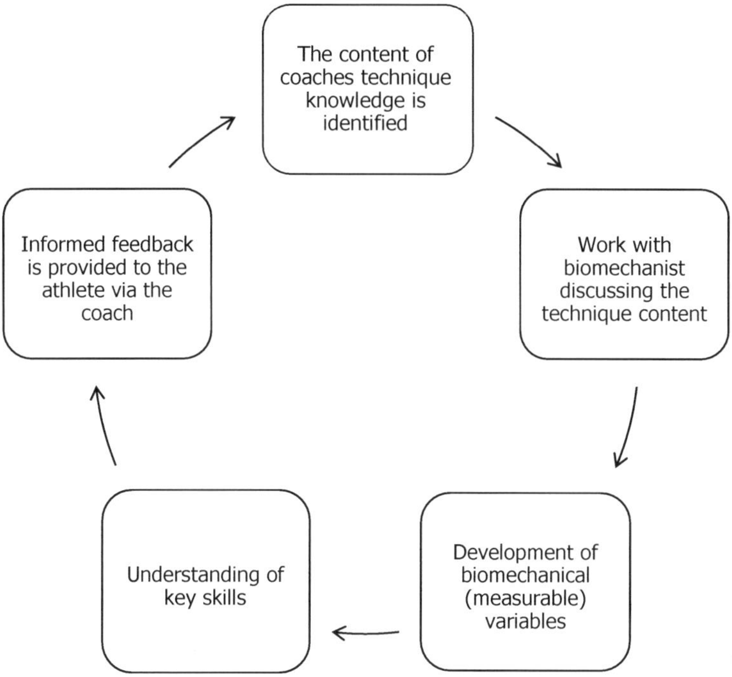

Figure 2.3: A diagrammatic representation of the biomechanics-coaching interface (Kerwin and Irwin, 2008).

One notion which has developed as a result of the relationship between coaches and sports biomechanists is known as the coaching-biomechanics interface (Kerwin and Irwin, 2008). This is a notion which describes a continuous process by which biomechanics can inform coaches about the performance of their athletes by relating the underlying biomechanical parameters to coaching information (Kerwin and Irwin, 2008, Smith et al., 2015), as depicted in Figure 2.3. Coaches' knowledge regarding the athlete's technique is combined with a sports biomechanist's knowledge of technique and the principles of biomechanics to determine measurable biomechanical parameters that are directly related to successful performance (Kerwin and Irwin, 2008). This information may enable and enhance understanding of the technique and potential progression drills as well as optimise the performance of the athlete following feedback given to the athlete via the coach (Kerwin and Irwin, 2008, Smith et al., 2015). This process is thought to aid coaches in a number of ways, including the development of: coaches understanding of technical skills; coaching and training practices; the evolution of technique skills (e.g. high jump, gymnastics skills in book); and optimising performance (Coleman, 1999, Kerwin and Irwin, 2008). Its purpose is to emphasise the importance of understanding biomechanics in relation to coaching and its use as a tool to be used critically by coaches by bridging the gap between biomechanics and coaching (Kerwin and Irwin, 2008).

The key factor in this process is communication between the biomechanist, coach and athlete (Kerwin and Irwin, 2008). Successful examples of the coaching-biomechanics interface and communication between sports biomechanics, coaches and athletes is evident in the literature:

- Irwin and Kerwin (2005) established forty-nine different skill progressions in the long swing in gymnastics which enabled coaches to rank the skill in terms of difficulty and select the appropriate skill progressions dependent on the athlete's capability.
- Thompson et al. (2009) analysed the technique characteristics that seven expert coaches associate with good sprint running technique. Four key technical parameters were identified (posture, hip position, ground contact and arm action) and compared to existing biomechanical literature, which highlighted discrepancies between the literature and coaches' knowledge.
- Elliott and Bartlett (2006) attempted to develop the technical performance of the tennis serve and identified the importance of internal rotation for the tennis serve and the body positions required to achieve the ideal internal rotation and racquet speed in the tennis serve. In another study, Elliott and Bartlett (2006) also used biomechanical principles to understand the biomechanics underpinning the javelin throw using computer simulations and investigated these in a practical context using top UK Javelin throwers. These pieces of research identified the need for understanding of biomechanics principles to enable effective feedback in terms of differences between good and bad performances.
- Smith et al. (2015) utilised the biomechanics-coaching interface to identify which key technical parameters sixteen high level golf coaches associated with a successful golf swing. A total of

five intrinsically linked key technical parameters were identified (posture, body rotation, arm and wrist action, sequential movement and body segments and club motion) and compared to existing biomechanical literature which also highlighted discrepancies between the literature and coaches' knowledge.

- Kerwin and Irwin (2008) highlighted that the athletic sprinter Darren Campbell worked alongside sports biomechanists in order to improve his sprint start technique for the 2000 and 2004 Olympic Games. Together they established the most effective sprint start technique for Darren as well as a theory regarding the relationship between thrust mechanics and a successful sprint start.
- Judge et al. (2008) attempted to integrate biomechanical analysis by using video analysis as part of a coaching system in the development of an athlete in the women's hammer throw. The study attempted to develop the understanding of the basic elements of the hammer throw technique in an effort to improve performance. It resulted in the athlete participating within the study producing an American record of 73.87m in the women's hammer throw in 2005, as well as progress in bridging the gap between sports biomechanics and coaching.

These examples highlight the potential contributions the discipline of biomechanics can make to coaching and sport performance. Although sports biomechanists and coaches appear to have the common goal of performance improvement in mind, their approaches differ and they tend to quantify and analyse athletes' movements in different ways (Federolf et al., 2014). As a result, literature from as early as the 1970's has highlighted the difficulties in bringing together the interests of these two groups (Martindale and Nash, 2013, Williams and Kendall, 2007). A number of factors identified in the literature in relation to these differences include:

- Conflicting terminology and language: A critical problem is the different languages and terminology spoken by sport scientists and coaches (Knudson, 2007).
- Research focus: The application of research into practice is an essential part of coaching development. Spinks (1997) drew attention to differences between the focus of sports science research projects and what coaches 'think' they need to know to be better coaches. Currently, the general consenus is that the transfer of sport science knowledge to coaching is poor and researchers are criticised for failing to ask the relevant questions and their findings being too difficult to apply (Williams

and Kendall, 2007). Sport scientists normally pursue research problems in the context of their own discipline or in areas where coaches do not need help. Alternatively, a coach needs to solve a problem specific to an individual or group of athletes, which might call for solutions that are multidisciplinary in nature (Martindale and Nash, 2013).

- Sources of information: Researchers are often judged by their peers in terms of the quality of their publications and status of the journals in which they publish. This has implications for coaches as an audience, given that they do not tend to read peer-reviewed scientific journals and are more likely to read sport periodicals and multi-disciplinary journals (Martindale and Nash, 2013). This may be due to a lack of time or limited access to sport science journals. Instead coaches have been noted to value experience and practical knowledge acquiried from participation in sport and from other coaches rather than knowledge gained from sports science research (Williams and Kendall, 2007).
- Funding limitations: While the availability of sport science support is on the increase, funding to provide sport science support is still relatively new for many sports and more often than not, it is aimed at elite level coaches and teams, leaving the use of sport science for most coaches to second hand and ad-hoc means (Martindale and Nash, 2013).
- Perceptions: Differing perceptions of both coaches and sport scientists have been noted in the literature in terms of their opinion of the opposite group of individuals and their roles (Martindale and Nash, 2013). Elite coaches have been noted to only listen to those sport scientists who could demonstrate a thorough understanding of the sports with which they worked; they only view knowledge as practical if it applies to their activities as a coach. Other coaches have been noted to see sport science as a useful and applicable part of coach education, highlighting the variation within this group of individuals in terms of understanding sport science and application (Martindale and Nash, 2013). Enabling an understanding of the qualities valued by coaches and sport scientists might be of assistance in establishing a productive working relationship between these two groups.
- Knowledge/Understanding: Research shows integration and application of knowledge involves a much deeper understanding than memorisation (Knudson, 2007).

As some literature has found good congruence between elite coaches and researchers in terms of research, although this focus was on the Australian institute of sport (Martindale and Nash, 2013; Williams and Kendall, 2007), some authors have attempted to bridge this gap and problem. Bishop (2008) developed an applied research model for sport science to help guide the research process to overcome these issues. However, its application in the literature has not been noted. The International Society of Biomechanics in Sports (ISBS) was also founded for this purpose and to provide knowledge for coaches; however, the extent to which that purpose is being achieved remains less than forecast (Sanders, 2015, personal communication). More work needs to be done to facilitate this transfer of knowledge effectively between sports biomechanists and coaches and it has been suggested that better communication and education between these two disciplines using the coaching-biomechanics interface may add applied value to this process (Kerwin and Irwin, 2008, Martindale and Nash, 2013).

Understanding how sports scientsts can support coaching practice is a task for both groups of individuals. Although there is much sport science research, research cannot keep up with the many changes in technique and equiqpment in the range of disciplines in sport. Thus, communication between these two groups is essential to ensure research is applied and relevant to the coaching process and athlete development (Knudson, 2007). Regardless of the method, further understanding and approaches are needed to begin bridging the existing gap between these two groups of individuals. This provides a valuable link between qualitative and quantitative approaches to technique analysis and a potential area in which research into how the gap between sport scientists and coaches can be bridged.

2.3.4. Feedback

As technique analysis is only effective if it can be applied to help improve the performance of athletes, one factor which is vital in this process is feedback. One of the main roles of feedback is to provide information for the performance (Guadagnoli et al., 2002, Schmidt and Lee, 1999). This provides the learner with information about how performance may be improved as well as motivating the athlete towards that goal. This is needed as many athletes cannot evaluate errors on their own (Guadagnoli et al., 2002). Part of the focus of the current literature is on the information coaches receive during observation and monitoring of fatigue during training. Therefore to inform and interpret the results of the present research, and compare them to other literature, some understanding of feedback is required.

The influence of feedback on motor performance and learning has been a popular topic of investigation in the motor learning domain since the 1960s (Magill, 2004, Phillips et al., 2013). The term feedback refers to performance-related information that the learner receives during and after performing the task, and there are two general types of feedback: intrinsic and extrinsic feedback (Pérez et al., 2009). Intrinsic feedback, also known as inherent feedback, is characterised by the sportsperson receiving the information "in real time" through different sensory mechanisms created as a result of performing a movement and which allows him/her to self-regulate movement and/or adapt the execution of the motor task to the model image, as in Figure 2.4 (Pérez et al., 2009, Zatoń and Szczepan, 2012).

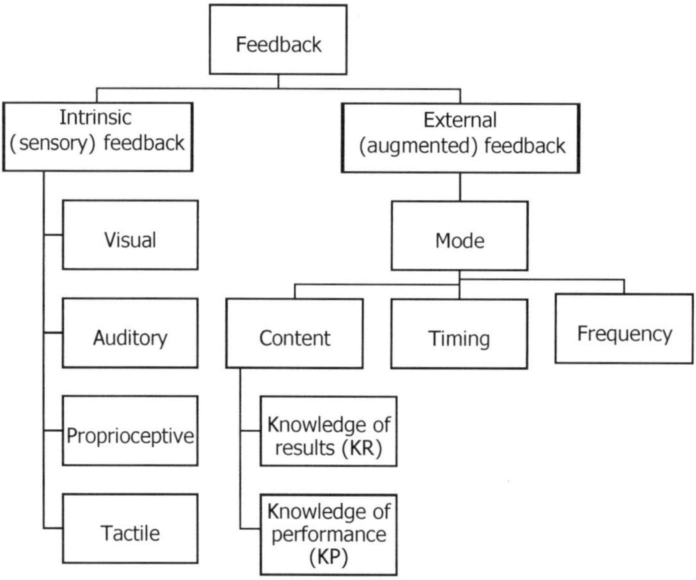

Figure 2.4: The various types of feedback available to an athlete and useable by a coach. Adapted from (Phillips et al., 2013).

Extrinsic feedback is the name given to the feedback that supplements intrinsic feedback, or increased feedback information, and refers to information from external sources such as the coach, another individual, or a video camera (Zatoń and Szczepan, 2012). Extrinsic feedback can be classified in a number of ways, as shown in Figure 2.4. Firstly, by the manner in which it is supplied, such as verbal or non-verbal or using external equipment (Hebert and Landin, 1994). Secondly, the content of the information provided such as the knowledge of the results and/or performance (Schmidt and Lee, 1999). Thirdly, the timing of feedback, such as concurrently while the movement is being carried out, terminally as soon as the movement stops, or provided with delay (Zatoń and Szczepan, 2012).

Feedback is seen as important in sport performance because repeating incorrect movement patterns consolidates erroneous movement patterns during training. The higher the number of incorrect repetitions, the more automatic the error and the more difficult it becomes to correct.

Immediate verbal feedback after occurrence of the error during training has a substantial influence on the process of automating appropriate movement patterns to optimise performance in competition. As a result, research has tended to focus on theory and concepts relating to feedback including its content, frequency or attentional focus (Wulf, 2007, Wulf and Shea, 2004). Guidelines for augmented feedback usage in relation to the frequency of use, timing of delivery and content are numerous, as are the associated examples of translation from theory into practice (Phillips et al., 2013). However, the majority of these evidence-based feedback guidelines are derived from laboratory-based experiments relying on single-joint movements performed by novices over short intervention periods, with any experimental measurement being focused on knowledge of results or outcome feedback (Baudry et al., 2006b). These experimental designs bear little resemblance to the applied sport environment. Further, they lack an appropriate rationale underpinning the selection and provision of specific movement patterning (kinematic), physiological and performance outcome information (Wulf and Shea, 2004).

In a sporting discipline such as swimming, the feedback that swimmers receive plays a key role in the technical performance, in both the learning process (Salmoni et al., 1984) and in the process of improving technique. The intrinsic feedback to the swimmer is always present, since it comes from the information that the nervous system receives through different receptors of the human body (Latash, 2008). Due to the environmental conditions that surround the swimmer (water, humidity, open space, etc.), there is limited or reduced use for electronic systems or instruments to give feedback (Pérez et al., 2009). For this reason, the extrinsic feedback that the swimmer receives is mostly verbal or with gestures supplied by the coach on the poolside providing information about technical performance (Pérez et al., 2009). As well as informing swimmers about the results (e.g. swim time), extrinsic feedback also has other implicit functions, as pointed out by Newell (1976), such as: as a guide to learning, an associative function, and a motivating or incentive function. These functions do not exclude each other but can be present at the same time, thus increasing the probability that the movement pattern may be modified appropriately to improve performance (Pérez et al., 2009).

Certain technological developments have enabled further feedback tools for coaches during the training process, including: pace-maker lights, dual-media images (Vezos et al., 2007), or speedometers (Seifert and Chollet, 2005). However, apart from the environmental problems that surround the swimmer, technology developments have also posed several problems, including the apparent 'need' to use or obtain such equipment, the lack of tools to store any data obtained from these devices, and the difficulties of applying these methods to training situations with multiple athletes (Phillips et al., 2013). The majority of this technology-driven research has lacked a suitable research design to accurately account for any genuine learning effects, and consequently this has not been systematically established (Phillips et al., 2013). Anderson (2010) has identified a 'triptych conundrum' associated with augmented feedback in biomechanical applications, and has argued that feedback technology must be: accurate and relevant; appropriately timed and delivered; and decipherable by the athlete. Nonetheless, research in this area continues to focus on what the technology is capable of doing, rather than being driven by establishing guidelines about how it should be used (Phillips et al., 2013).

2.3.4.1. Summary

The coaching process is not flawed, but the observation and analysis phases of the process are limited. Many great coaches are able to identify changes required to influence performance, but even the best can make wrong decisions. The role of feedback is central in the performance improvement process, and by inference, so is the need for accuracy and precision of such feedback (Hughes and Franks, 2007). Although there is considerable empirical evidence of correct methods for coaches and teachers in providing feedback to athletes generally (Hughes and Franks, 2007), less is known about the effectiveness of instruction.

Providing evidence-based recommendations for the use of feedback technology and method is difficult, partly as a result of limitations of current research methods and traditional laboratory-based work (Phillips et al., 2013). Working collaboratively and in an inter-disciplinary manner between the disciplines of sport science, researchers may be able to systematically examine, investigate and address pertinent research areas (Phillips et al., 2013).

2.3.5. The Role of Video Analysis in Technique Analysis

One of the main tools used to aid feedback and technique analysis is motion or video analysis. Traditionally, cinematography (using cine cameras) was used; however, as this is now rarely used, it will not be considered within this review (Bartlett, 2007). Over the last thirty years, the development and use of video-based technologies within sport performance environments has greatly increased (Carling et al., 2014, Mackenzie and Cushion, 2013), as well as a parallel advancement in computer technology. Such developments have made this type of equipment relatively low cost, portable and more accessible to those involved in sport performance (Kerwin and Irwin, 2008). As a result there has been a rapid increase in the use of video in coaching as a tool to provide extrinsic feedback to both the coach and athlete and augment traditional coaching methods (Hughes and Franks, 2007, Wilson, 2008).

Video is believed to be especially useful in terms of allowing coaches to generate detailed quantitative and qualitative analyses of individual and collective sporting performances, as well as offering great potential in facilitating the provision of feedback to athletes (Booroff et al., 2015, Nelson and Groom, 2012). As a result it is associated with a number of advantages that may not be accessible using traditional coaching methods. These include:

- Video provides an objective record of a performance which may not be seen readily by coaches observing the movement by the naked eye in real time. In any sporting situation, it is difficult for coaches to notice and remember all the key events which occur, meaning the information collected may not be reliable (Booroff et al., 2015, Wilson, 2008).
- Video also provides the coach with an extra tool when working with an athlete which can be used by the coach to 'see more' (Booroff et al., 2015). Using several video cameras and taking video from the front, side and back, the coach can have access to several different views of the athlete's performance.
- The coach may examine a performance repeatedly and use video playback options, such as slow motion, freeze frame or replay to allow the coach to review and analyse the athlete's entire performance after the event, leaving the coach to focus on a particular aspect of the athlete's performance during the actual event (Booroff et al., 2015). These tools enable the enhancement of feedback during video replays.

- The camera can be placed in areas where the coach can't go. For example, underwater video of swimming provides an indication of how the above-water positions and timing of technique can be influenced by the underwater stroke actions (Booroff et al., 2015).
- For athletes, viewing videos of themselves has been shown to improve performance (Thow et al., 2012); however, this may only be of any good if the coach and athlete can make comparisons of their performance with the 'ideal' role model or a 'before and after' comparison of themselves to identify areas of development.

Video is mostly recognized as an appropriate medium for obtaining information about performance (Liebermann et al., 2002). As coaches are at the forefront of providing feedback to athletes on a daily basis, the potential application of this tool will only be reached if coaches lead the way. It is within their power to integrate feedback-based technologies into their training protocols.

Video feedback can be used to address issues raised by both the coach and athlete and the advice offered must be strongly grounded in biomechanical principles rather than using a good performer as a role model (Knudson, 2007). The feedback format and presentation provided should match the level of the coach and athlete to avoid information overload (Hodges and Franks, 2002). Unfortunately, our understanding of the ways in which coaches and performance analysts utilise technology is basic. No research has been conducted in swimming on this topic and much of the available performance analysis literature has either focused on how to utilise various video-based technologies to effectively examine the performances of athletes and teams or, instead, the generation of detailed, descriptive accounts of the technical, tactical, biomechanical and physiological dynamics of sporting performance (McGarry et al., 2013, Peters and O'Donoghue, 2013). Further research is needed to understand the uses of video by coaches and performance analysts, as well as their effectiveness in applied sporting situations, such as training.

2.3.5.1. 2-D video analysis methods

Performance analysis is now acknowledged as an aid to performance enhancement at all levels. It is about creating a valid and reliable record of performance by means of systematic observations that can be analysed with an aim of improving performance (Phillips et al., 2013). From video, analysts may extract both qualitative and quantitative information. The two disciplines of coaching and sport science may use similar methods to collect data for analysis but the main thing they both have in common is measuring the observation during or after an event to quantify performance in an accurate, reliable and valid way (Phillips et al., 2013, Pike, 2008). Without some means of recording, even the most experienced coach cannot achieve this type of observation. As a result of technology development, more and more accessible video analysis software systems are becoming available for use in sports.

One such system manufacturer is Dartfish Ltd. which provides a range of commercially available software programmes that can incorporate feedback into a video of a motor performance (Pike, 2008). The software is primarily marketed to coaches, but is also designed for other applications. Video can be imported into the programme either directly from a camcorder or by uploading video file formats (Pike, 2008). Once imported, video files can be split, trimmed, duplicated and converted to different formats to accommodate the user's needs. Qualitative and quantitative tools are available to analyse movement and provide visual augmentations to video feedback, allowing a coach to provide visually augmented video feedback immediately following performance (Pike, 2008). The software's tools include measurements of angles, distances and time, as well as various methods of highlighting certain locations or events. This endows the user with the ability to control what information is presented.

Hodges et al. (2007; p. 542) specifically indicate that Dartfish software *"provides practitioners with a viable method to manipulate access to relevant...information in the field setting"*. The product is already in widespread use with athletes at all levels in many sports (Bartoli et al., 2004, Baudry et al., 2006a, Hars and Calmels, 2007, Hodges et al., 2007, Williams and Hodges, 2005). It has also been used in several studies in swimming to identify the timing of arm and leg phases (Alberty et al., 2008, Alberty et al., 2009, Alberty et al., 2011). However, no literature has been identified pertaining to its use in terms of coaches' feedback. Analysis based on accurate observation and recall is a key tool for improving future performance and software such as Dartfish may be the tool to bringing coaches and sport scientists closer together (Bishop, 2008a).

2.4. SUMMARY OF LITERATURE REVIEW

While the relationship between kinematic technique factors and the resulting stroking parameters and swim speed is clear, the impact of fatigue and the role of these factors during training remains unknown, particularly in breaststroke swimming. The literature reveals some evidence that fatigue can have an impact on swimmers' ability to maintain stroking parameters during high-intensity training-like sets. However, no research has investigated the effects of fatigue during high-intensity sets on kinematic technical actions. Most of the conclusions from the literature have been derived from laboratorybased research, using quantitative research approaches which are often not applicable in a training environment, and thus not useable or practical for coaches.

Traditionally, augmented feedback in sports took the form of verbal information communicated by coaches or sports scientists, based on their perception of the performance or via simple devices such as a stopwatch (Phillips et al., 2013). However, as technology has evolved, specific measurement tools such as video analysis, now provide a myriad of extremely precise information and feedback opportunities for the coach. Nevertheless, despite the obvious potential of such tools to provide feedback to facilitate skill learning and performance, there has been a lack of systematic research on the most effective manner to utilize such technologies to provide augmented feedback in the sport performance environment (Phillips et al., 2013). In addition, no research has been conducted into the application of 2-D video methods by coaches in a training environment to monitor fatigue.

Thus it remains unclear whether high-intensity training has any influence on a swimmer's capacity to sustain optimal kinematic technical actions, nor whether coaches can observe these variables using traditional coaching methods or 2-D video analysis methods. This issue is addressed in the present research by attempting to assess coaches' capacity to observe technical changes in technique (as a result of fatigue) during training and investigating the potential implications of using the 2-D video analysis software Dartfish during training.

Chapter 2: Summary

What was already known about this topic?

1. A large gap exists in the literature on the following topics: o The effects of fatigue on technique during training in swimming.
    - Coaches' current knowledge and perceptions of fatigue during training in swimming.
    - The coaching practices used to monitor fatigue during training in swimming.
    - The applicability of video analysis methods into coaching practice during training in swimming.
2. A gap currently exists between sport scientists and coaches in terms of research, knowledge and the methods they utilise.

What new information does this chapter provide?

- Kinematic technical actions play a key role in determining propulsive and resistive forces during swimming, and therefore influence the stroking parameters and swimming speed. However, little research has been conducted to analyse these kinematic parameters during training scenarios.
- No research has been conducted to analyse the effects of fatigue on kinematic technical variables during training-like scenarios in breaststroke.
- There are differences in approaches used by coaches and sport scientists, including analysis methods and perceptions of sport science. Ways of 'bridging the gap' and removing communication barriers need to be implemented.
- 2-D video analysis and software may be a viable tool in beginning to bridge the gap between sport scientists and coaches by its application during training.

3

An Investigation into the Quality of Data from Dartfish 2-D Video Analysis in Swimming

3.1. INTRODUCTION

It is clear that 2-dimensional (2-D) video analysis is being increasingly used as a coaching and feedback tool in a range of sports (Eltoukhy et al., 2012, Garhammer and Newton, 2013, Liebermann et al., 2002). As the success of coaching feedback depends, among other things, on the effectiveness of the observation and its analysis, it is extremely important that any information collected is objective, unbiased, and accurate (Hughes and Franks, 2007). Many coaches, teams and organisations have undertaken video-based analysis in an effort to supplement their direct observational skills during training and competition, and enhance their feedback in an effort to improve their athletes' performances (Kerwin and Irwin, 2008, Pearce, 2005).

The increased use of 2-D video analysis methods by coaches may be for several reasons. Firstly, the progress of video technology has advanced its application and feedback capacity. Video now has a number of functions, including slow motion and frame-byframe playback allowing feedback to be presented quickly and easily (Wang and Parameswaran, 2004). Secondly, 2-D analysis equipment and software have become increasingly cost-effective and user-friendly (Bertram et al., 2007). Commercial software (e.g. Dartfish, Silicon Coach, or Kinovea) suitable for this type of analysis can run on a standard laptop or tablet and only require a low-cost digital video camera for input (Kerwin and Irwin, 2008). Conversely, although 3 dimensional (3-D) motion capture, e.g. Vicon (OMG Plc, Oxford, UK) or Qualysis (Qualysis, Sävebalden, Sweden) is considered the gold standard method for motion analysis (Eltoukhy et al., 2012, Garhammer and Newton, 2013), it tends to be expensive and difficult to use. It can require multiple cameras, advanced knowledge and training, and analysis procedures can be time consuming, delaying the feedback to athletes (Eltoukhy et al., 2012, Kirmizibayrak et al., 2011). Due to time limitations, 3-D analysis is often regarded as not feasible nor applicable for use by coaches during training (Kirmizibayrak et al., 2011, Nash, 2008). The technological progression of 2-D video analysis methods has enabled an alternative method and opportunity for coaches to monitor their athletes quickly and effectively (Bertram et al., 2007).

In order to maximise the feedback obtained from video, it is vital to ensure that the video recording is of a high quality (Hughes and Franks, 2007). Throughout this thesis quality will be defined as '*the standard of something as measured against other things of a similar kind; the degree of excellence of something*' (Oxford Dictionary). The collection of video data and video analysis involves many procedures, which if completed incorrectly can induce errors resulting in low quality video data. This error is known as measurement error, defined as the difference between the observed value of a measure and the true value (Hopkins, 2000). There are two types of measurement error; systematic and random. Systematic errors are predictable, occur in one direction and are consistently over- or under-estimating the true score (Hopkins, 2004, Portney and Watkins, 2000). Random errors on the other hand occur due to chance (Portney and Watkins, 2000). Measurement errors can arise from the equipment (movement of the camera, lens distortion due to camera's optical system, precision limits in the digitisation process related to coordinate resolution or computer round-off errors, refraction from the environment) and the operator (inaccuracies in the calibration measurement or digitiser errors during analysis due to movement of markers) (Bartlett, 2007). Furthermore, the processes and methods of 2-D video analysis can make certain measurements more susceptible to errors:

- Timing measurements/key time points: These can be limited to the frame rate (sampling frequency). Frame rate is '*the number of complete full images captured per second*' (Payton and Bartlett, 2007; p. 13). A frame rate of 25Hz, for example, captures a complete image every 0.04s. Thus any actions between these points or frames, e.g. hand entry, could be missed.
- Angle and displacement measurements: Errors can arise due to the misalignment of the camera, perspective error (due to

subject or calibration object being out of photographic plane and operator errors of judgement) and parallax error (viewing actions away from the optical axis of the camera and across the plane of motion) (Bartlett, 2007, Brewin and Kerwin, 2003).

If the information or data obtained from video data has errors and is of a poor standard, this could influence the feedback provided to an athlete and consequently be detrimental rather than beneficial to the athlete's performance. It is therefore imperative that all errors should be minimised by good experimental procedure, and any remaining errors are identified or removed (Bartlett, 2007).

Over many years, reviews in biomechanics have emphasised the importance of ensuring that measurements made as part of athlete support or research are of a high standard and errors are quantified and minimised (Atkinson and Nevill, 1998, Payton and Bartlett, 2007). This can be achieved by establishing the validity, reliability, accuracy and precision of a measure, although various definitions for these terms are often used interchangeably in the literature. For the purpose of this study they will be defined as:

- Validity: This is the '*ability of the measurement tool to reflect what it is designed to measure and the agreement between the value of a measurement and its true value*' (Atkinson and Nevill, 1998, Hopkins, 2000). There are four types of validity: face validity (instrument appears to test what it is supposed to test); content validity (an instrument adequately samples content that define the variable being measured); criterion-related validity (outcomes of an instrument or target test, can be used as a substitute measure for a gold standard criterion test); and construct validity (ability to measure the degree an instrument reflects theoretical components) (Portney and Watkins, 2000).
- Reliability: The '*consistency or reproducibility of a measure*' (Atkinson and Nevill, 1998, Hopkins, 2000). Estimates of reliability can vary depending on the type of reliability being analysed. This study will focus on test-retest and rater reliability only. Test-retest reliability is defined as the capacity of an instrument to measure a variable with consistency (Atkinson and Nevill, 1998). Rater reliability can be defined as intra-rater, which is the stability of data collected and analysed by one individual across more than 2 trials, or inter-rater which is the variation between two or more raters (Portney and Watkins, 2000). This research will look at intra-rater reliability only.
- Accuracy: This is quantified as the difference between a true value and an observed value (Payton and Bartlett, 2007).

Accuracy is deemed different to validity as it is shown to quantify the bias in a measure, caused by measurement error, not the agreement between the true and observed values (Payton and Bartlett, 2007).

- Precision: This is defined as the differences between an observed and an expected mean value (Payton and Bartlett, 2007).

Validity and reliability are characteristics all instruments have to a degree and both are fundamental to the process of research (Portney and Watkins, 2000). Without reliability, rational conclusions cannot be drawn from data with certainty and without validity, inferences cannot be drawn from data nor generalised (Portney and Watkins, 2000). It must be noted that measures can be reliable without being valid and thus assessing both of these factors is important to address whether a method can discriminate, evaluate or predict relevant outcomes (Portney and Watkins, 2000). The precision and accuracy of a measurement can also relate to its validity and reliability. If validity and reliability are poor it can reduce the precision and accuracy of a single measurement or the ability to track changes or characterise relationships over time (Hopkins, 2000, Payton and Bartlett, 2007). Although precision and accuracy are sometimes confused, they are distinct qualities and when both are high, the measurement is deemed to have little error (Payton and Bartlett, 2007). With any new approach it is necessary to establish proof of its capacity and differentiate any errors in the measurement (Coutts et al., 2014). Quantifying these factors in any measurement is important so that the results of any analysis can be concluded with certainty and are not simply a result of measurement error (Bartlett, 2007). This is especially important in performance sport where the differences or margins investigated can be very small and there is a need to provide the athlete with meaningful information (Coutts et al., 2014, Hopkins, 2000). By using methods of a high standard, coaches will be able to interpret and provide comprehensive feedback to their athletes with a degree of certainty. This can have a vital role in the monitoring of athletes during the training process.

Despite the importance of assessing and detailing the standard of data provided from 2D video analysis, the majority of research in this area has only taken place in the past decade. This research has concentrated on test-retest reliability and concurrent validity of 2-D video analysis methods in functional screening tests with non-athletic populations. As this is not the main scope of this thesis, this will be covered only briefly here. The functional screening tests have included various step or jump movements, with recordings taken from both the front (Hollman et al., 2009, Miller and Callister, 2009, Munro et al.,

2012) and sagittal planes (Norris and Olson, 2011). Other studies have also been completed on the reliability and validity of measurements of lateral trunk motion of jumping actions (Dingenen et al., 2014), and spine kinematics (Mier, 2011). Mytton et al. (2013) noted the importance of camera angles and positioning to the reliability of measures and the potential of increased errors if this was done incorrectly. These papers highlighted that 2-D video analysis methods have high concurrent validity, intra-tester and test-retest reliability when measuring lower limb static and dynamic angle measures. Although important, these results are only applicable to similar actions and measures which are non-sporting actions.

Although video analysis using 2-D methods is deemed popular amongst coaches, research in sport in this area has lagged behind those trends in current coaching practice (Bertram et al., 2007). Those studies using 2-D video analysis have mainly analysed kinematic parameters during sports performance and include:

- Angle, distance and timing measurements at key instances throughout a hang power clean (Rucci and Tomporowski, 2010).
- Analysing absolute trunk, thigh and relative knee angles in the trail leg in curling (Kraemer, 2009).
- Trunk and lower extremity variables in volleyball (Parsons and Alexander, 2012).
- Applying the timing tool to public performance videos in 1500m running (Mytton et al., 2013).

Despite the use of 2-D video methods, only one group of authors, Rucci and Tomporowski (2010), established the reliability of their measures between testers. Rucci and Tomporowski (2010) established that all the measures had a high inter-tester reliability using an intra-class correlation (ICC), with the angle measures ranging $r = 0.94 - 1.0$, the distance measures ranging $r = 0.96 – 0.98$ and the time values ranged $r = 0.98 - 1.0$. Although video analysis has been used for many years, the majority of this research has taken place within the past five years, highlighting the recent use of these methods in sport science research and the existing gap between sport science research and current coaching practices (Nash and Sproule, 2011). Further research is needed not only to bridge this gap, but also to enhance the applicability of this method. In order to apply and use 2-D video analysis methods in a coaching environment it is important that coaches can utilise these methods with confidence and achieve valid, reliable and precise measures. This is often achieved through the comparison of a measure to a criterion standard; however, research continues to focus on simple measures during simple actions, even in a sporting context and very few of these papers stated any detail regarding the reliability, validity or accuracy of these measurements.

One of the most practical and objective approaches to validity testing is based on the ability of one test to predict the results obtained by another test. One test is often compared with a gold standard or criterion measure that is already established (Portney and Watkins, 2000). In video analysis, the gold standard method is 3-D analysis, however, the comparison of 2-D video analysis data to other methods is scarce in the literature. Eltoukhy et al. (2012) identified that for a simple squat motion the difference in angle measures of Dartfish (Dartfish Ltd, Fribourg, Switzerland) in comparison to 3-D Vicon software (OMG Plc, Oxford, UK) data was 4.7 - 6.9% for the knee angle, and 13.119.1% for the ankle angle. Eltoukhy et al. (2012) suggested the differences were due to the different ways in which the two systems measure angles. Garhammer and Newton (2013) compared the measurement of vertical displacement, horizontal displacement, and vertical velocity during the performance of a squat. These measures taken by Ariel Performance Analysis System (APAS, Ariel Dynamics Inc, CA, USA) (the criterion values) were compared to four 2-D analysis software packages: Dartfish version 4.5.2 (Dartfish Ltd, Fribourg, Switzerland), Kinovea version 0.8.15 (Kinovea Association, France, www.kinovea,org), Logger Pro version 3.8.4 (Vernier Software and Technology, Oregon, USA, and Tracker version 4.62 (www.opensourcephysics, org). Garhammer and Newton (2013) found that, in comparison to the measures taken by APAS, Dartfish showed the highest error in the measure of horizontal displacement (4%) and vertical barbell velocity (9%). Part of this error may be linked to the simple scaling often used in 2-D video analysis and the potential errors which can arise from this (Brewin and Kerwin, 2003, Payton and Bartlett, 2007). The remaining types of software were thought to be equivalent in accuracy to APAS in the measurement of vertical displacement, horizontal displacement, and vertical velocity as the values for each measurement were within 12% or less. This literature suggests that for certain measurements, some 2-D video analysis software can produce data as reliable and valid as methods which have already been established; however, discrepancies do exist between these methods for aspects such as: types of measurement; joint location; and scaling approaches. Despite these errors and issues, 2-D video analysis software is still considered useable for athletic support and frequently used by coaches and sport scientists to monitor athletes (Miller and Callister, 2009).

In addition, kinematic measures from 2-D video analysis systems have been used as criterion values against which the same measures from other equipment have been compared, including: Infrared Optojump systems (Microgate Co., Bolzano, Italy) (Balsalobre-Fernández et al., 2014); Brower audio sensor (Draper, UT, USA) an Omega timing systems (Swiss timing, Corgemont, Switzerland) (Haugen et al., 2012); and GPS accelerometer technology (minimax S4, Catapult Sports, Melbourne, Australia) (Beanland et al., 2014). Previous research investigating the standard of 2-D video data for sport analysis has again focused only on simple, single-plane sport actions in laboratory controlled environments. More research is needed to apply this analysis method in a coaching environment effectively.

In addition to the potential source of errors in 2-D data from land-based activities, the data collected using 2-D video analysis methods in swimming research can be subject to additional errors caused by factors such as: distortion from light reflection of the water; poor visibility under the water and from the swimmer's actions; the 3-D nature of the movements in swimming (Callaway et al., 2009). The research which has pertained to using 2-D analysis methods in swimming only provided detail of the accuracy of Dartfish software in the measurement of stroke phase durations. This was specified by each piece of research as 0.02s, operating at 25Hz, and explained as being due to the interlacing capacity of the software, but no other measures of validity or reliability were taken. Leblanc et al. (2009) completed angle measures of the thigh and distances from a frontal perspective, however again there was no detail regarding whether parallax or perspective errors were quantified during this process (Bartlett, 2007). It is important, particularly when videoing underwater that these issues are taken into account and managed to minimise errors and aid the production of high quality video data.

Other studies using 2-D video analysis in swimming have focused upon analysing certain technical aspects, such as stroking parameters or timing in front-crawl swimming. Schnitzler et al. (2009) used Dartfish Pro version 4.0 (Dartfish Ltd, Fribourg, Switzerland) to look at stroking parameters and stroke phase timing in front-crawl 400m swimming. Similar factors have also been analysed by a group of researchers (Alberty et al., 2008, Alberty et al., 2011), again looking at front-crawl 400m paced swimming whilst controlling the pace or swimming to exhaustion. Similarly, the software was only used to analyse stroking parameters or stroke phase timings. Leblanc et al. (2009) adapted the methods used by Alberty and co-workers for use in the assessment of breaststroke technique factors but also limited its use to stroke phase timings and coordination. No measure of reliability, validity or accuracy was assessed in these studies.

To summarise, research on the use of 2-D video analysis in swimming is sparse with current studies restricted to general stroke parameters. To date, no studies have utilised the software to analyse specific technical markers in swimming, particularly in breaststroke. More research is needed to assess the use of 2-D video analysis in swimming over a variety of disciplines, technical factors and measures in order to assess its applicability for use by coaches. This is to ensure coaches are providing comprehensive feedback to their swimmers and maximising the use of video to monitor athletes during training.

3.1.1. Purpose of the Study

Information on the standard of data produced in sports involving multi-plane actions through the use of 2-D video methods is essential due to this method's accessibility and application in coaches' feedback methods in many sports at present. This information is critical to ensure this method can provide high quality data and aid coaches (Justham et al., 2008, Slawson et al., 2008). To date, no research has investigated the attributes of data provided from 2-D video analysis, specifically Dartfish, for technical actions while swimming. Therefore, in order to successfully provide answers to the primary research questions of this thesis, the standard of kinematic technical measures using Dartfish must be established for a swimming action and underwater video analysis. To achieve this, this chapter aims to:

- Establish criterion values and correction factors.
- Assess the validity of the measurement of angles and distances using Dartfish software.
- Determine the accuracy and precision of the measurement of angles and distances, in-and out-of-plane of the camera, specifically relating to predetermined technical variables in breaststroke swimming using Dartfish software.
- Ascertain the reliability of measuring the predetermined breaststroke technical variables using Dartfish software.
- Confirm a series of breaststroke technical variables to be used as markers within this thesis based upon the standard of their measurement.

3.2. METHODS

This section details the methods used to assess the validity, accuracy, precision and reliability of the distance and angle measures. To clarify this explanation, some results and outcomes will be provided in this section. The remaining results will follow in Section 3.3.

3.2.1. Criterion Measures and The Minimisation of Errors

3.2.1.1. The T-frame design and determination of criterion measures

To determine criterion measures a T-shaped frame was constructed. The frame consisted of two perpendicular metal bars shaped in the form of a 'T'. Each metal bar contained three plastic balls (diameter 3cm) securely fastened along the bar at set intervals. The six balls and the centre joint of the two bars provided seven analysis points, as shown in Figure 3.1.

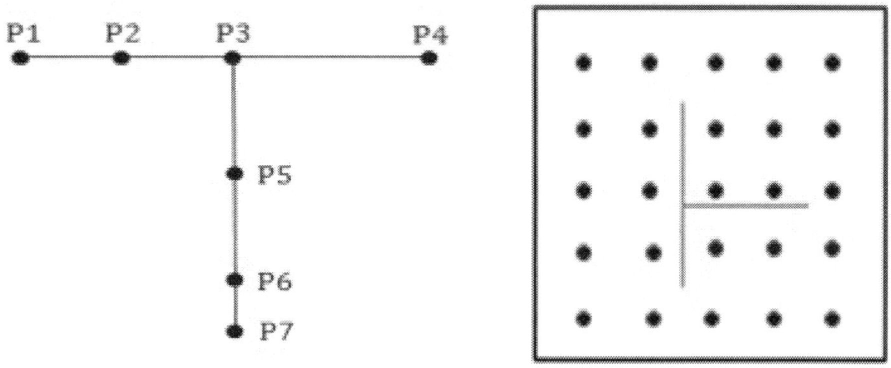

Figure 3.1: The land T-shaped frame and 5x5 point grid.

To determine the x-y coordinates of the seven analysis points on the land T-shaped frame, two video recordings were taken. The first recording, used to determine the calibration data, was of a 5x5 grid drawn upon the surface of a digitiser tablet (TDS LC series II, TDS Nuomonics, Blackburn, England) (See Figure 3.1). These data were obtained using a method described by Kelly (2007). The grid lines were spaced at 0.25m for the horizontal and 0.23m for the vertical, i.e. 1m x 0.9m (Kelly, 2007). This was the maximum size that could be completed on the tablet. Each of the 25 intersections was marked with a black circular sticker, 2cm in diameter. The second video recording was of the T-shaped frame positioned flat against the surface of a digitiser tablet, in the centre of the 5x5 grid, and rotated to the left, so that the top of the T-frame was vertical and the tail of the 'T' was pointing to the right, as shown in Figure 3.1. For both video recordings, a digital camera (Panasonic VC-100, Panasonic Corporation, Osaka, Japan) was positioned perpendicular to the centre of the digitiser tablet, at a distance of 15m and a height of 1.32m. Each recording lasted a total of

10s at 50Hz, using 1/250 second shutter speed. This was the highest shutter speed that could be obtained with the available lighting.

Each recording was analysed using the Ariel Performance Analysis System (APAS, Ariel dynamics, CA, USA). The first video was trimmed to 30 frames and each grid point manually digitised using the methods described by Kelly (2007). The second video was trimmed to 50 frames and the centre of each of the seven analysis points on the T-shaped frame was also manually digitised. Each set of data was exported to Microsoft Excel, with each of the 25 grid points and the seven analysis points of the T-shaped frame obtaining an x-y coordinate from each frame digitised. There was no Z coordinate. The mean x-y value of each grid point, determined from 30 frames, was then used as the calibration data in the analysis of the T-frame. The mean x-y value of each T-shaped frame point, determined from 50 frames, was taken as the criterion x-y values for each analysis point (See Appendix 1). The digitising error, as indicated by the standard deviation, was less than 0.2mm for the grid point and T-shaped frame. The mean x-y coordinate values were taken as the final values for both the calibration data and the T-shaped frame as they were deemed to reflect the 'true' value as indicated by the Central Limit Theorem (Field, 2009). This theorem indicates that any errors thought to be from this process would be normally distributed and any random errors would cancel each other out (assuming no systematic error) due to the large sample size (greater than 30) and likelihood that the digitising errors were random (Field, 2009, Hopkins, 2000).

Lastly, the x-y coordinates for each of the seven points on the T-shaped frame were used to determine the values of the criterion distances and angles, using Pythagoras Theorem, and are presented in Table 3.1. These measurements attempted to include those with an internal reference or relative to an external point (i.e. joint or segment measures) in an effort to represent all potential errors (Kelly, 2007).

Table 3.1: The criterion distance and angle values for the T-shaped frame. (m) = metres, (°) = degrees.

Criterion distance values (m)				**Criterion angle values (°)**			
Horizontal measures		**Vertical measures**		**Horizontal measures**		**Vertical measures**	
^{L}1-2	0.14	^{L}3-5	0.25	^{L}3-4-5	49.4	^{L}3-5-4	40.6
^{L}1-3	0.28	^{L}3-6	0.50	^{L}3-4-6	67.0	^{L}3-6-4	23.0
^{L}1-4	0.49	^{L}3-7	0.57	^{L}3-4-7	69.6	^{L}3-7-4	20.4
^{L}2-3	0.13	^{L}5-6	0.25	^{L}3-1-5	41.9	^{L}3-5-1	48.1

Criterion distance values (m)				Criterion angle values (°)			
Horizontal measures		Vertical measures		Horizontal measures		Vertical measures	
^{L}2-4	0.35	^{L}5-7	0.32	^{L}3-1-6	61.1	^{L}3-6-1	28.9
^{L}3-4	0.21	^{L}6-7	0.07	^{L}3-1-7	64.2	^{L}3-7-1	25.8
				^{L}3-2-5	61.9	^{L}3-5-2	28.1
				^{L}3-2-6	75.2	^{L}3-6-2	14.8
				^{L}3-2-7	76.9	^{L}3-7-2	13.1

To collect data in the water using this tool, the bottom of the T-shaped frame was attached through the centre of a float (a kickboard) to a metal rod. The metal rod runs the length of the base of the float and a short distance in front of the float. An additional float (a pull buoy) was attached to the front end of the metal rod to ensure the frame and float remained level on the water surface during use (See Figure 3.2 below).

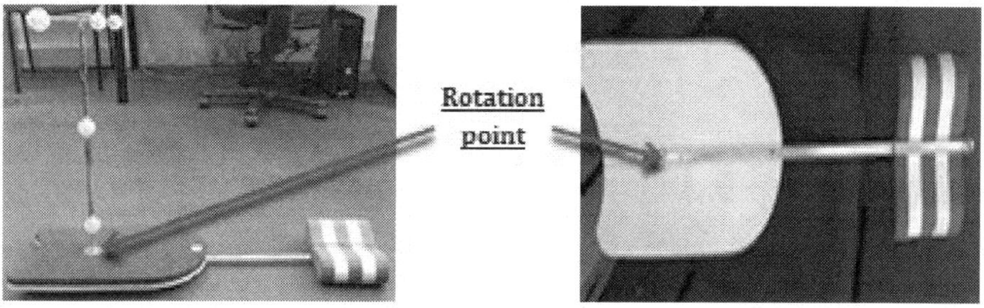

Figure 3.2: The T-shaped frame and rotation point.

Due to the attachment of the T-shaped frame to the metal rod on the float, the T-shaped frame could be rotated in increments of ± 30°. This enabled the assessment of the standard of measures taken both 'in-plane' and 'out-of-plane' of a camera perspective which was important to ensure all potential measurements were assessed and are representative of the actions which could take place while swimming.

3.2.1.2. Correction factor calculations for distance measures

Data collected in the swimming pool environment in this and subsequent chapters, was calibrated and analysed using Dartifish motion analysis software, version 6.0 (Dartfish Pro Suite motion analysis software, Dartfish Ltd, Fribourg, Switzerland). To calibrate a recording using Dartfish software requires the measurement of a known distance, using the distance tool. Within this thesis, this consisted of the measurement of a distance of 1m from two calibration lines placed over the path of action, recorded from a side view perspective (See

Figure 3.3). Each calibration line ranged from wall to wall, at a length of 25m, and balls (diameter 2cm) were securely fastened at 1m intervals throughout the length of each line. However, any measurements of distance taken outside the calibration may result in parallax and perspective errors in the video data and must be corrected to reflect the true distance values (Bartlett, 2007). Two such corrections were used in this thesis in an effort to reduce these errors:

1. Side view camera correction: The optical axis of the side view camera was aligned with the middle of the pool, defined as zero (See Figure 3.4). Distance measurements from the optical axis taken on the swimming pool wall, 1m further from the camera than the calibration line, reflected distance values different from those on the path of action. The offset of these two distances must be established and corrected to reflect the true distance point on the path of action throughout the side view camera's entire field of view.
2. Front-view camera correction: The distance measures became distorted as a swimmer moved closer or further away from the front-view camera position, and thus closer or further from the calibration point. If not corrected, this would result in over- or under-estimation of the distance value.

To calculate the offset and correct for the side view camera errors, the difference in horizontal distance points between the measures on the wall and that expected on the path of action was determined using known distance values and Pythagoras Theorem (see Figure 3.3). The distance from the left of the optical axis was defined as positive and to the right as negative.

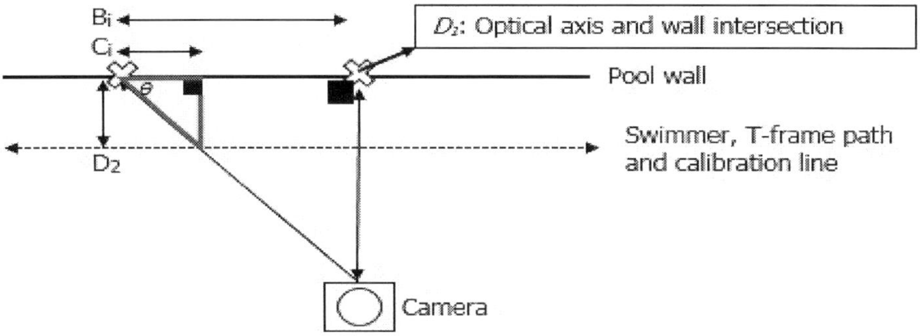

Figure 3.3: Calculation of the side view correction factor. This image shows the relevant distance and angle calculations to calculate the distance offset between the pool wall and path of action from the optical axis of the centre side view camera.

This resulted in the following two calculations to determine the angle from the wall to the camera (Θ) and the resulting distance offset (Ci):

$\Theta = \tan_{-1} (D_1) / (B_i)$ Equation 1

Where D_1 is the optical axis, a distance of 13.25m from the wall to Camera 1, and B_i is the horizontal distance from the optical axis.

$C_i = \tan(\Theta) \times D_2$ Equation 2

Where (Θ) is the angle calculated in Equation 1 and D_2 is the distance from the path of action to the wall (1.31m).

Table 3.2: The side view correction factor. The raw data of the offset value, corrected distance and validation of this measurement. (m) = metres.

	Correction		Validation
Distance from optical axis (m)	**Offset (m)**	**Corrected distance from optical axis (m)**	**Error in distance correction (m)**
-3.00	-0.30	3.30	0.01
-2.50	-0.25	2.75	0.00
-2.00	-0.20	2.20	-0.01
-1.50	-0.15	1.65	-0.01
-1.00	-0.10	1.10	0.00
-0.50	-0.05	0.55	0.00
12.50	0.00	0.00	0.00
0.50	0.05	0.45	0.00
1.00	0.10	0.90	-0.01
1.50	0.15	1.35	0.00
2.00	0.20	1.80	0.00
2.50	0.25	2.25	0.00
3.00	0.30	2.70	-0.01
3.50	0.35	3.15	0.01

The full corrected distance points on the path of action are available in Table 3.2. The corrected distance points were validated by analysing the video of the calibration from the underwater side view camera perspective using Dartifish Pro analysis software, version 6.0. Using the Dartfish measurement tool, the distance from the optical axis (0) to each distance was measured on the calibration line, using the wall and its markings as a guidance tool. This was compared to the corrected distance on the swim path, measured from the mid-pool point (0 or 12.5m). As shown in Table 3.2 above, there was a difference of 0.01m or less between the corrected distance on the path of action and the Dartfish measurement. The corrected values were used to define the distance of the swimmer from the centre optical axis when measurements were taken.

The data to determine the front-view correction scale factor was established using a twodimensional video recording method (Bartlett, 2007). Data were collected in a single lane at the wall side of a swimming pool 25 x 13m, with a depth of 1.8 - 2.0m, at a fixed water temperature of 29°C. Four video cameras (Elmo-PTC-450c colour, Elmo Co. Ltd, Aichi, Japan) were used, recording at a sample rate of 25Hz. Although a recording sample rate of 50Hz would have been preferable, this was the highest that could be obtained with the available underwater video system. Each camera was stationary and positioned as shown in Figure 3.4, modified from Thow (2010). Camera 1 was located at the centre of the pool, at a depth below the water of 0.5m and perpendicular to the line of movement. Camera's 2 and 3 were located at 0m and 25m respectively, both at a depth of 1.5m below the water and in line with the path of action. Camera's 2 and 3 were located at a lower depth so as to be out of the way of the swimmer's turn and swim performance. Camera 4 was also located at the centre of the pool, at a height of 0.25m above the water level and directly above Camera 1. All cameras were synchronised (frame synchroniser), (Data video TBC-5000, Data Video Technologies Co. Ltd, Taipai, Taiwan) with a synch pulse generated by a black burst generator (Kramer 810B, Kramer Electronics, Ltd, Buckinghamshire, UK). Time codes were used to identify aligned frames in each video (Vorte TC generator, Vortex Communications Ltd, London, UK). The video data were stored as AVI files until analysis. This video recording was arranged to replicate the data collection set up to be used in the following chapter (Chapter 4).

Prior to data collection, a calibration recording was taken. This consisted of the placement of two calibration lines; one line was attached at water level height and the other at a depth of 1.5m in the middle of the allocated analysis swim lane and directly over the lane T-line situated on the pool floor. This was due to the fact the technical measures would be taking place from the water surface to depths below the water, depending on the measurement. Each line was tightened to ensure it was taught and recorded twice, for ten seconds. To ensure the line tension was consistent, three distances were taken (the vertical distance between each calibration line and the horizontal distance from the first marker to the wall on each calibration line) for every calibration recording.

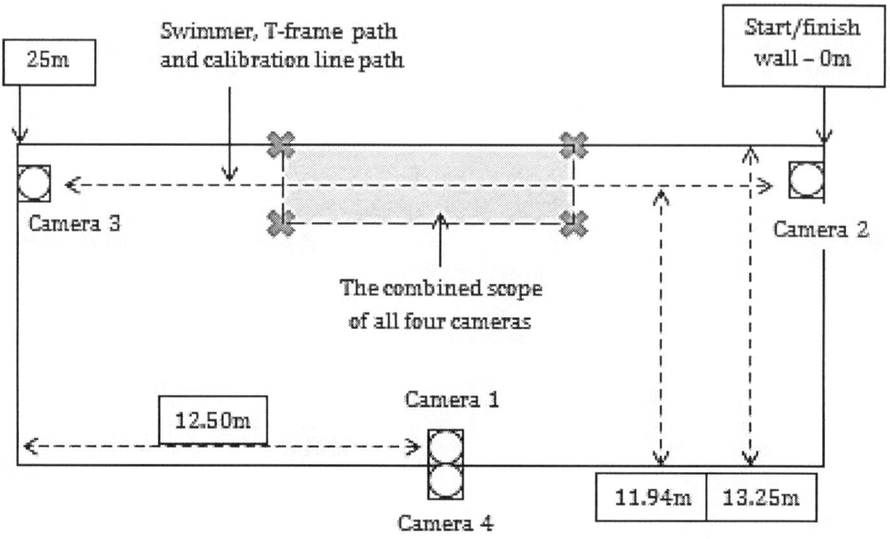

Figure 3.4: The camera layout for video analysis in the swimming pool. (m) = metres.

The water T-shaped frame was then towed through the water, the length of the pool (25m), using a rope. It followed the calibration line, passing through each camera perspective (both above and below the water). The T- frame was positioned at 90° relative to the below water, front-view camera and throughout the recording, stopped and held (to ensure it was flat and still on the water) at 0.5m intervals between 9.5 and 16.5m of the pool for each recording.

A 1m distance between each ball was measured from the side below camera view of the calibration recording for both the surface and 1.5m deep calibration line. This measure was used as a distance calibration. A measurement was taken from both calibration lines to ensure that the calibration distance was the same regardless of the difference in depth.

The distance between analysis points 1 and 4 on the T-frame (See Figure 3.1) was then measured on the under-water front-view recording every 0.5m between 9.5m and 15.5m using the distance tool on the Dartfish software. The Dartfish measurements were compared to the criterion distance (L_{1-4} = 49.03cm), using Equation 3, to give a scale factor value for each of the fifteen values (See Table 3.3).

Scale factor = measured distance/criterion distance Equation 3.

Table 3.3: Correction scale factor for front view camera distance measures. (m) = metres

Corrected distance (m)	Dartfish Measure of $L_{1\text{-}4}$ (m)	Scale factor
9.80	0.41	0.84
10.3	0.45	0.92
10.7	0.49	1.00
11.2	0.53	1.08
11.6	0.57	1.16
12.1	0.6	1.22
12.5	0.64	1.31
13.0	0.68	1.39
13.4	0.72	1.47
13.9	0.76	1.55
14.3	0.8	1.63
14.8	0.84	1.71
15.2	0.88	1.80

To determine whether the correction scale factors could be applied accurately and assess the relationship, the fifteen scale factor values were plotted against the corrected distance measures for each swimming direction, both described in the previous section, using a scatter chart (Portney and Watkins, 2000). A linear trend line was added to each scatter plot to determine the line equation, the correlation value (r), and the standard error of estimate (SEE). These were used to assess the strength of the association and the accuracy of predictions from the scale factor (Portney and Watkins, 2000).

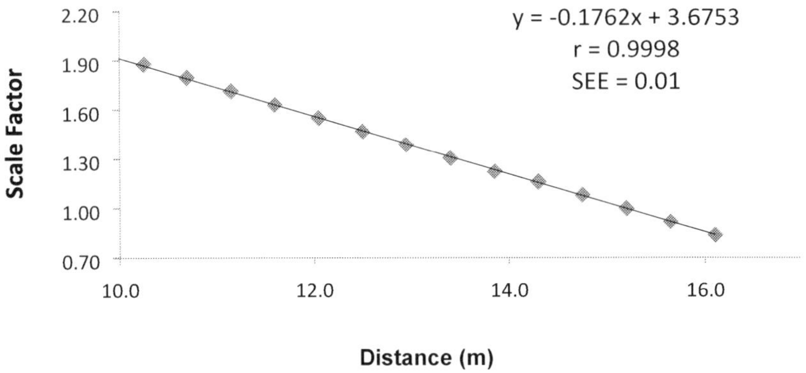

Figure 3.5: The validity of the scale factor for swim laps 1 and 3.

There was a strong, statistically significant association between the scale factor value and actual distance, r = 0.9998, $p < 0.05$, as shown in Figure 3.5 (Field, 2009). The low SEE values (0.01) also indicate low variability in these values. Therefore, these scale factor values could be used to correct the front view measures and produce accurate values (Portney and Watkins, 2000).

From this information the regression line equation was used to estimate the correction scale factor for distance points every 0.25m. Each front-view distance measurement taken within this thesis was then divided by the relevant correction scale factor; depending on the distance from the object to the camera at the point the variable was measured. The distance of the object from the camera was found using the side above (Camera 4) and below (Camera 1) views.

3.2.2. The Validity of Angle and Distance Measures in a Water Environment using Dartfish Software

To assess the validity of the measurement of angles and distances, in a water environment, using Dartfish, the criterion values were compared with the same measurements taken using Dartfish. To achieve this, the video recording set-up and calibration recording was the same as that described in section 3.2.4. Following this, the water T-shaped frame was then towed continuously through the water, the length of the pool (25m), using a rope. It followed the path of the swimmer and calibration line, passing through the field of view of each camera. The T-shaped frame was positioned at 0°, 90° and 30° (relative to the front view camera) and a recording taken (for both above and below the water) of the T-shaped frame being towed through the water at each angled position. This was to assess the validity of all potential measurements, whether taken while the object was in- or out-of-plane relative to each camera, and ensure that these measures were representative of the actions that could take place while swimming.

Following these recordings, the data were calibrated, again using the methods described previously, and the criterion angles and distances established in section 3.2.1, were measured on the T-shaped frame using Dartfish software. The measures were taken in the centre of the pool at 1m intervals, between 9.5m and 15.5m, in the combined scope of all the cameras (See Figure 3.4). This was completed for each of the four camera views with the T-frame at each of the set angles. The Dartfish measurements were then compared to the criterion values calculated in section 3.2.1. This was completed for the distances and angles individually using the following method: firstly, the mean of the

Dartfish measures, taken every 1.0m (seven values), was calculated for each distance and angle measure; secondly, the mean value and criterion value were compared using a paired t-test, with a confidence level of $p < 0.05$ accepted as statistically significant, to assess if there were statistically significant differences between the two values. A paired t-test was used as there was only one independent variable (each measurement) and due to the use of only one analyser (Field, 2009). The null hypothesis was defined as showing no statistically significant differences and the data were tested for normality using a Shapiro-Wilks test. Any data that were not normally distributed were assessed nonparametrically using a Wilcoxon signed-rank test (Field, 2009).

Finally, to assess if there was a relationship between the Dartfish measures and criterion values, a Pearson's Correlation Coefficient was used (Portney and Watkins, 2000). The validity of the measures was assessed by the standard error of estimate (SEE), the r value from the correlation and the R^2 coefficient of determination value. These methods were again used to assess the strength of the association and the accuracy of predictions from it (Portney and Watkins, 2000). In addition, the coefficient of determination was used to further assess the association between the variables and show the importance of any effects (Field, 2009). The statistical analysis was completed using the Statistical Package for Social Sciences, SPSS (version 19.0, IBM UK Limited, Portsmouth, UK).

3.2.3. The Accuracy and Precision of Angle and Distance Measures in a Water Environment using Dartfish Software

As all measures have error it is important to establish the accuracy, precision and error in a measurement. To ensure the Dartfish measures used in this thesis are accurate and precise, two factors were investigated; (1) the accuracy and precision of the measures of distance and angles by Dartfish; (2) the accuracy and precision of the kinematic technical measures in the following chapter, based on the error in the angle and distance measures.

The criterion angle and distance values on the T-shaped frame (described in section 3.2.1) were measured repeatedly on seven different occasions (separated by 24 hours) when the T-shaped frame was positioned at 0°, 90° and 30°. The following four calculations were then completed on this data;

- The relative mean and standard deviation of all seven values was calculated for each criterion distance and angle measurement on the T-frame. The standard deviation determined the precision of these measures by quantifying the deviation between a criterion and observed value and is actually an indication of a measure's inaccuracy (Payton and Bartlett, 2007).

- For each of the seven measures taken for all distance and angular measurements, the Dartfish measure was subtracted from the criterion value to determine the error in its calculation.
- To calculate the accuracy, the root mean square (RMSQ) of the error values from the seven measures were established for each criterion measure (Payton and Bartlett, 2007). The RMSQ of the differences between the criterion and observed value was used to indicate accuracy as it is the most conservative criterion (Payton and Bartlett, 2007).

The data from these calculations were then applied to the kinematic technique measures proposed for analysis in Chapter 4. The technical variables selected were measured due to their importance during the performance of breaststroke technique and by applying hydrodynamic principles of propulsion and resistance in conjunction with literature relating to breaststroke technique faults (Maglischo, 2003). A detailed description of each variable and how it was measured is provided in Appendix 2. The assessment of the accuracy and precision of the technique measurements, which were made up of combinations of angle/distance measures varying from in-plane to out-of-plane, was completed to ensure that the technical variables measured in future chapters within this thesis were also accurate and precise. To determine the accuracy and precision of each technical variable, the following three concepts, as well as each variable's measurement process, were considered:

- Identify the number of joint/limbs/segments, over which measures took place, for the technical variable i.e. two limbs combined, one limb.
- Whether the technical variable is measured as distances, angles or both.
- Whether each part of the measure was taken in-or out-of-plane of the camera (or both).

The accuracy (RMSQ) and precision (SD) of the T-shaped frame criterion values were applied to the three concepts above. The relative, potential criterion accuracy and precision measures were summed and averaged for each technical variable. If more than one average accuracy and precision value was needed for a technical variable (e.g. it covered more than one joint segment or used distance and angle measures), the two average values were summed. This was to provide the maximum potential estimate of accuracy and precision for the measurement of each technical variable, from the Tshaped frame data.

3.2.4. The Reliability of Angle and Distance Measures on Swimmers in a Water Environment using Dartfish Software

Eighteen breaststroke swimmers, 9 male and 9 female, (age 18.4 ± 2.5 years, body mass 67.7 ± 10.3 kg, height 175.8 ± 10.2cm) currently competing at national level or higher, participated in this study. All participants were currently participating in a full-time periodised training program and their regular weekly training schedule consisted of 16.5 ± 2.5h water-based training and 4.7 ± 1.9h dry land training. The participants' 100m breaststroke personal-best times were 64.1 ± 5.3s and 71.8 ± 4.1s for the male and female participants, respectively. These times, expressed as a percentage, were within 15% of the 2012 Senior and Junior, short course British records for 100m breaststroke,

(calculated as the percentage of BR= $\text{time}_{\text{subjects}} / \text{time}_{\text{British record}} \times 100$, (Leblanc et al., 2009)). This was 109.6% ± 4.3 and 107.4% ± 4.5 for the male and female participants, respectively. The sample was limited to elite, national level or higher swimmers to increase the likelihood that technique characteristics and patterns would be well established and consistent (Pyne et al., 2004, Nikodelis et al., 2005). For additional participant information, see Table 3.4.

Participants were excluded from the study if they: were not a specialist breaststroke swimmer currently competing at national level or higher; were with injury or in the process of recovering from an injury; had a limiting medical condition at the time of the study which made it unsafe for them to participate. Prior to the test session, participants were informed of the purpose of the study, the experimental protocol, and provided written informed consent. For those individuals under 16 years of age, the child's parent or guardian's written informed consent was also required before participation. The research and paperwork were approved by the University of Edinburgh Ethics Committee, see Appendix 3.

Table 3.4: A description of the participant characteristics.

Athlete	Age (yrs)	Weight (kg)	Height (cm)	Experience (yrs)	100-m Brst P.B (SS:MS)	British record (%)
1	20	67.95	170.8	12	68.5	103.9
2	18	71.3	178.3	10	62.2	107.4
3	19	59.7	164.5	12	67.8	102.8

Athlete	Age (yrs)	Weight (kg)	Height (cm)	Experience (yrs)	100-m Brst P.B (SS:MS)	British record (%)
4	22	68.5	181.1	10	69.1	104.8
5	19	78.3	180.0	10	62.0	107.1
6	18	56.85	171.0	12	71.0	107.7
7	20	75.9	186.1	10	67.0	115.7
8	22	73.5	178.4	15	66.7	101.1
9	19	77.4	185.2	8	61.0	105.4
10	19	79.5	182.4	10	63.8	110.2
11	15	66.4	183.5	5	69.0	109.8
12	17	70.3	174.3	8	68.0	117.4
13	15	50.0	162.0	5	78.0	113.9
14	19	78.5	184.3	12	61.0	105.3
15	17	60.3	163.0	8	74.0	112.2
16	14	48.75	160.0	10	76.0	111.0
17	16	55.6	163.0	7	75.0	109.5
18	23	80.0	196.0	8	62.7	108.2
Mean	18.4	67.7	175.8	9.6	67.9	108.5
S.D	2.5	10.3	10.2	2.6	5.3	4.4

To minimise the effects that confounding variables (including: heavy/overtraining (Arroyo-Toledo et al., 2013, Richmonda et al., 2015); caffeine (Burke, 2008); alcohol (supplements) (Shirreffs and Maughan, 2006); dehydration (Sawka et al., 2007); lack of sleep (Sargent et al., 2014)) could have on performance, each participant was asked to continue their normal dietary habits, ensure they were well rested and hydrated prior to their test session and avoid the previously mentioned confounding variables prior to their testing session.

The data were collected in the same pool and conditions described in section 3.2.1.2 Before completing the test session, each participant was marked with black waterproof actors' paint applied with a 3cm circular sponge at nine anatomical landmarks and joint centres, on both sides of the body (Bartlett, 2007, Wren et al., 2008). The markers include the vertex (centre of the head at the highest point); C7 (on the anterior at the Adams apple); wrist (styloid process of ulna), elbow (midway along line joining medial and lateral epicondyles of humerus), shoulder (head of humerus), hip (greater trochanter of femur), knee (lateral epicondyle of femur), ankle (lateral malleolus of fibula), Xiphoid (bottom of the sternum). The pubis was also used during the video analysis but it was not marked. These markings were used to aid the identification of the anatomical landmarks and body midline during video analysis.

Each participant completed a warm-up lasting thirty minutes, to ensure they were prepared for the test session and to reduce the risk of injury (Balilionis et al., 2012). The warm-up was standardised to avoid differences in fatigue levels prior to testing and consisted of traditional warm-up activities, including both aerobic and race pace swimming, swim drills, and stretching (Balilionis et al., 2012, Bishop, 2003). After a five minute period of passive rest, participants completed 5x25m's breaststroke swim at 100m race pace on three minutes, from a push start. The correct swim pace was checked using the split times against each individual's 100m split (100m time divided by four). After each 25m swim, the participants swam back to the start end of the pool and were required to rest passively until the next 25m swim. Both active and passive rest were used as the recovery method in accordance with research which highlights that both forms of recovery combined, following vigorous activity, are more beneficial in terms of aiding the recovery process (Toubekis et al., 2005). It was also similar to training sets used in swimming (Maglischo, 2003). To ensure each swimmer was performing at maximum effort at the initiation of the test set, the 5x25m swims had to be performed within 5% of the individual's personal best (Thow, 2010). This was achieved successfully by all the participating swimmers, emphasising the consistent high effort and motivation of participants in this research. This protocol was the same as that to be used in Chapter 4.

All performances were recorded as per the method of Thow (2010) with the modification described in section 3.2.1.2. Throughout the 5x25m swims, all the technical variables were analysed over two stroke cycles, using Dartifish Pro analysis software, version 6.0, and the methods described in Appendix 2. Each camera perspective allowed the measurement of certain variables.

To assess the reliability of the investigator's digitising technique on the aforementioned technical variables, three swimmers' results were selected randomly. This involved placing the swimmers in alphabetical order by surname and numbering them, a dice was then thrown and the number identified selected the athlete's to be used for analysis (Thomas et al., 2011). For each of the three athletes, each technical variable was measured over one stroke cycle, 10 times, separated by twenty-four hours. The standard deviation and 95% confidence interval of error of the pooled data from repeated digitising of all three individuals, was used to determine the digitising reliability. These statistics were calculated using Microsoft Office Excel 2010 software. The test-retest trial reliability was assessed for each variable using the average of the two stroke cycles per length, over the 5x25m swims. This gave five entry points for seventeen individuals. This was assessed using the typical error, shift in mean and intra-class correlations (ICC) of the 5x25m swims by all individuals and calculated using Hopkins' website (2002). The ICC was interpreted according to Munro et al. (2012) as; poor <.40, fair .40 to .70, good .70 to .90, and excellent >.90.

To assess whether any fatigue effects occurred within the 5x25m trials, a repeated measures ANOVA, with the lap number as the factor, was used to evaluate whether any technical variables differed statistically significantly in magnitude relative to each other. Each lap represented the average of two stroke cycles. This was repeated for each technical variable. No statistically significant differences were identified. Therefore, it could be considered that the 5x25m protocol included sufficient rest and did not incur fatigue effects in the participants. This analysis was completed using the Statistical Package for Social Sciences, SPSS (version 19.0, IBM UK Limited, Portsmouth, UK).

3.2.5. The Exclusion of Technical Variables for Future Analysis

The validity and reliability of a test can have a high impact on the ability to track and monitor an athlete's performance (Hopkins, 2004). In sports performance, measures must be sensitive enough to detect changes in performance which may be very small yet still meaningful, particularly for athletic performance (Atkinson and Nevill, 1998). This is known as the smallest worthwhile change and is used to determine if a real change has occurred over time (Spencer et al., 2006). By taking account of the magnitude of the smallest worthwhile change and the error (or noise) in the test measure, it can be determined whether any changes in parameters are a result of measurement error or a result of performance changes (Hopkins, 2004, Pyne et al., 2006). If the error is less than the smallest worthwhile change, the measure is rated as 'good'. However, if the error is considerably greater than the smallest worthwhile change, then the measure is rated as 'marginal' or 'poor' (Spencer et al., 2006).

To determine the magnitude of a worthwhile change, the smallest worthwhile change was calculated for each technical variable as 0.2 (the default for the smallest worthwhile effect) of the between-swimmer variability in accordance with existing methods (Fulton et al., 2009, Hopkins, 2004, Pyne et al., 2006). The between-swimmer variability was determined from the 5x25m swim data. The smallest worthwhile change value for each technical variable was then compared to the calculated accuracy (RMSQ); precision (SD); the digitising error indicated by the standard deviation and 95% confidence intervals; the test-retest trial reliability indicated by the typical error, shift in mean and ICC data measured in sections 3.2.3 and 3.2.4, respectively. Only those variables which had error values less than the smallest worthwhile change and an ICC value of greater than 0.90 were to be included for further analysis in Chapter 4. This was to ensure that any observed changes were not likely to be due to measurement error, and that only those technical variables which could be measured accurately, precisely, reliably and validly, from a 2-D perspective, were included.

3.3. RESULTS

3.3.1. The Validity of Angle and Distance Measures in a Water Environment using Dartfish

Only the set of data for angles measured from the front below camera perspective, with the T-shaped frame positioned at 30°, was found to be not-normally distributed. No statistically significant differences were found in the Paired t-test comparisons (ranging p = 0.082 to 0.922). The Pearson correlation coefficients showed strong, statistically significant associations between the Dartfish software measures and criterion values, with r =0.98 or higher, $p < 0.001$. The R^2 value was also very high ($R^2 > 0.98$) for all variables apart from angles measured out-of-plane using the front below camera view, which still had a high value of $R^2 = 0.96$. To assess the accuracy of the relation between the criterion values and Dartfish software measurements, the standard error of estimate (SEE) was calculated. The SEE identified that the distance values measured using Dartfish software were very accurate at predicting the criterion values (Howell, 2012). The values of 0.01m or less indicated accurate and low error in the measurement of these variables in comparison to the criterion measures. A higher error was found for the angle values when measurements were taken out-of-plane of the optical axis of each camera perspective (i.e. 30°). This was particularly apparent in measures taken out-of-plane for the front below camera perspective. This indicates that the Dartfish software measurements had strong, statistically significant associations with the criterion values, low measurement errors and no statistically significant differences between these values for all cameras and perspectives, see Table 3.5 (Portney and Watkins, 2000). However, caution should be taken with angle measurements taken out of-plane of all three camera perspectives as, although the differences were not statistically significant, larger errors were found. The full statistical results from this study can be found in Appendix 4.

Although it is known in 2-D analysis that the camera should be positioned perpendicular to the action taking place (including height), due to certain limits of the studies reported in subsequent chapters, certain cameras were required to be placed at a depth of 1.5m. Therefore, to also ensure measures from this depth were accurate, additional recordings and measurements were taken with the camera system positioned at 0.25m below the surface to assess whether the difference in camera depth had an impact on the ability to measure distance and angle values accurately. Only two statistically significant differences between the different camera depths were found when distance measures were taken from the side view camera perspective (both above and below the water, p = 0.001 and 0.002, respectively) with the T-shaped frame angled out of frame at 30°. All other measures were not statistically significantly different when a raised or lowered camera position was used ($p > 0.05$). The full results can also be found in Appendix 4.

Table 3.5: Validity of criterion measures. SEE = standard error of estimate, ° = angles, (m) = metres, R = correlation, R^2= coefficient of determination.

		Criterion angle measures			**Criterion distance measures**		
	T-frame angle	**SEE (°)**	**R**	**R^2**	**SEE (m)**	**R**	**R^2**
Side below camera	0°	0.52	1.00	1.00	0.00	1.00	1.00
	90°	-	-	-	0.00	1.00	1.00
	30°	1.20	1.00	1.00	0.01	1.00	1.00
Side above camera	0°	0.62	1.00	1.00	0.01	1.00	1.00
	90°	-	-	-	0.00	1.00	1.00
	30°	1.05	1.00	1.00	0.01	1.00	1.00
Front below camera	0°	-	-	-	0.01	1.00	1.00
	90°	0.65	1.00	1.00	0.00	1.00	1.00
	30°	6.17	0.98	0.96	0.01	0.99	0.99

3.3.2. The Accuracy and Precision of Angle and Distance Measures in a Water Environment using Dartfish Software

The mean accuracy, precision and percentage error for all distance or angle measures in each camera perspective and T-shaped frame angle were calculated. A summary of those results is presented in Figures 3.6 and 3.7. The average values of precision and accuracy were similar for all camera perspectives and T-shaped frame positions at 0.01m, apart from the front below camera perspective, angled at 30°, which had a value of 0.2m for both precision and accuracy.

A similar pattern was found in the average angular precision and accuracy measures. The front below camera perspective, angled at 30°, had higher precision and accuracy values in comparison to the other camera measures. The higher these values were, the lower the accuracy and precision of the measurement. The precision and accuracy of the angle measurements was better when measures were taken in-plane than out-of-plane, as indicated by the higher results for all measures taken at an angle of 30°. In addition it was also apparent that the measures were more precise than accurate.

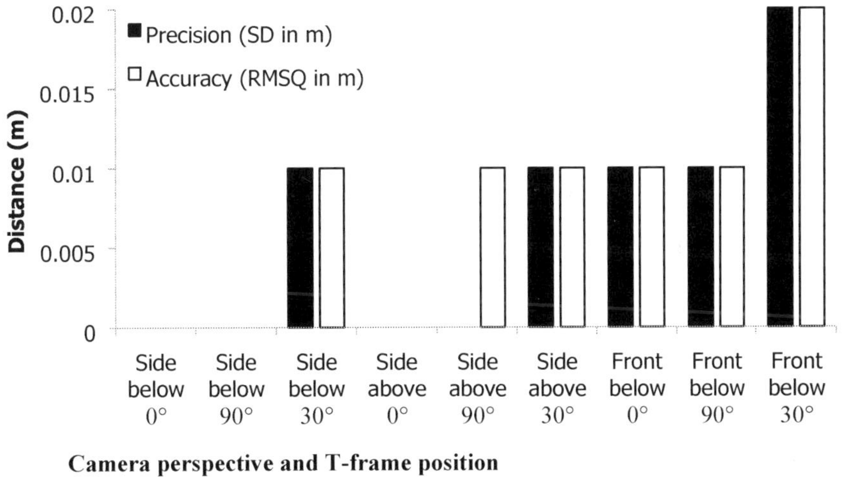

Figure 3.6: The precision and accuracy of Dartfish distance measures to the nearest 0.01m. This figure shows the precision and accuracy of each distance measure from each camera perspective and T-frame position. If no value, it indicates a value of 0. SD = standard deviation, RMSQ = root mean square error, (m) = metres.

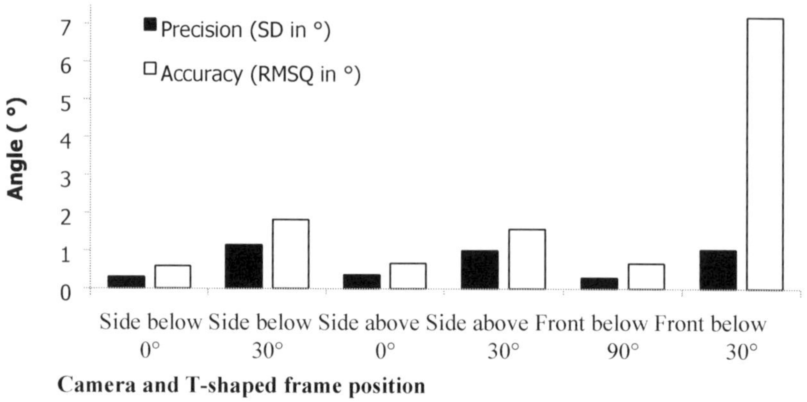

Figure 3.7: The precision and accuracy of Dartfish angle measures (°). This figure shows the precision and accuracy of each angle measure from each camera perspective and T-frame position. SD = standard deviation, RMSQ = root mean square error, (°) = angles.

This indicated that the worst error appeared to be for the front-view camera perspective, when both distance and angle measurements were taken out-of-plane of each camera perspective. The complete results of the accuracy and precision in the angular and distance measurements, from each of the three T-frame angles (0, 90 and 30 °) is in Appendix 5.

The precision (standard deviation) and accuracy (RMSQ) for each technical variable are presented in Table 3.6. The values for accuracy, precision and error are small for most variables. However, some variables should be considered carefully, namely; the frontview angular variables which had much higher values of precision, indicating that these measures were not as precise as angle measurements taken from the other camera perspective. Namely;

- Elbow angle at end of arm out-sweep
- Elbow angle at end of arm in-sweep
- Hip angle to vertical at the end of leg recovery
- Knee angle at the end of leg recovery
- Ankle angle at beginning of leg out-sweep
- Knee angle at beginning of leg out-sweep
- Knee angle at beginning of leg in-sweep

The technical variables measuring distance from the front view perspective also appeared to have a marginally larger accuracy and precision values compared to variables measured from a side above or below-water perspective. This may be due to a continued effect of the perspective/parallax errors created using a front-view camera perspective and the object of interest moving towards or away from the camera.

Table 3.6: The precision and accuracy values for each technical variable. This table shows the precision and accuracy of each technical variable. SD = standard deviation, RMSQ = root mean square error, (°) = angles, (m) = metres, (m/s) = metre per second, (s) = second.

Technical Variable			
Number	**Name**	**Precision (SD)**	**Accuracy (RMSQ)**
1	Horizontal alignment end arm rec (°)	0.29	0.62
2	Hip depth minimum (min) (m)	0.00	0.01
3	Hip depth minimum (max) (m)	0.00	0.01
4	Maximum foot displacement (m)	0.00	0.01
5	Elbow angle end arm rec (°)	1.36	2.60
6	Hip angle end arm rec (°)	1.36	2.60
7	Hip angle end leg rec (°)	1.36	2.60
8	Hip angle begin leg in-sweep (°)	1.36	2.60
9	Hip angle begin leg out-sweep (°)	1.36	2.60
10	Knee angle end arm rec (°)	2.42	3.34
11	Knee angle end leg rec (°)	2.14	3.94
12	Knee angle begin leg in-sweep (°)	2.14	3.94
13	Knee angle begin leg out-sweep (°)	2.14	3.94
14	Maximum hand displacement (m)	0.00	0.01
15	Head displacement at breathing (m)	0.00	0.00
16	Trunk angle during breathing (°)	0.34	0.67
17	Hand disp. end arm in-sweep (m)	0.03	0.03
18	Elbow angle end arm in-sweep (°)	2.06	15.22
19	Elbow angle end arm out-sweep (°)	2.06	15.22
20	Hand disp. end arm out-sweep (m)	0.03	0.03
21	Hip angle to vertical end leg rec (°)	0.96, 0.01	7.71, 0.01
22	Knee angle end leg rec (°)	1.92	15.45
23	Knee disp. end leg rec (m)	0.03	0.03
24	Feet disp. end leg rec (m)	0.03	0.03
25	Ankle angle begin leg out-sweep (°)	2.20	15.02
26	Knee angle begin leg out-sweep (°)	1.92	15.45
27	Knee angle begin leg in-sweep (°)	1.92	15.42
28	Knee disp. begin leg in-sweep (m)	0.03	0.03
29	Foot disp. begin leg in-sweep (m)	0.03	0.03
30	Arm phase timing (s)	0.04	0.04

Technical Variable			
Number	**Name**	**Precision (SD)**	**Accuracy (RMSQ)**
31	Leg phase timing (s)	0.04	0.04
32	Average velocity (m/s)	0.01, 0.04	0.01, 0.04
33	Stroke length (m)	0.01	0.01
34	Stroke rate (m and s)	0.01, 0.04	0.01, 0.04

3.3.3. The Reliability of Angle and Distance Measures in a Water Environment using Dartfish

Table 3.7 presents the results from the investigator's digitising technique, based on the same trial of three participants being analysed ten times. The 95% confidence intervals show the range in which the true value of that particular variable fell 95% of the time. The average standard deviation (SD) of all three individuals is considered small and acceptable for these variables.

The SD of the test-retest trial reliability were also considered small and acceptable for most variables. However, some variables need to be interpreted cautiously. The percentage typical error and shift in mean is also considered small in comparison to the mean. The intra-class correlation (ICC) for most variables is shown to be above 0.90 (a high correlation) apart from a series of timing variables (see Appendix 6 for full results).

Table 3.7: A summary of the reliability of the breaststroke technique variables. CI = confidence interval, SD = standard deviation, ICC = intraclass correlation, (m) = metres, (°) = degrees.

	Technique variable	**Digitising Error**			**Inter-trial reliability**		
			95% CI				
		SD	**Low 95%**	**High 95%**	**Typical error %**	**Shift in mean %**	**ICC**
1	Horizontal alignment end arm rec (°)	0.06	0.02	0.11	18.93	3.31	0.98
2	Hip depth minimum (min) (m)	0.00	0.00	0.01	3.16	0.87	0.94
3	Hip depth minimum (max) (m)	0.01	0.00	0.01	3.41	1.25	0.91
4	Maximum foot displacement (m)	0.01	0.00	0.01	1.38	0.22	0.99

	Technique variable	Digitising Error			Inter-trial reliability		
			95% CI				
		SD	Low 95%	High 95%	Typical error %	Shift in mean %	ICC
5	Elbow angle end arm rec (°)	0.44	0.12	0.75	0.16	0.06	0.99
6	Hip angle end arm rec (°)	0.35	0.10	0.59	0.16	0.03	1.00
7	Hip angle end leg rec (°)	0.33	0.09	0.56	0.22	0.04	1.00
8	Hip angle begin leg in-sweep (°)	0.24	0.07	0.40	0.23	0.04	1.00
9	Hip angle begin leg out-sweep (°)	0.60	0.17	1.02	0.29	0.09	1.00
10	Knee angle end arm rec (°)	0.43	0.12	0.74	0.18	0.03	1.00
11	Knee angle end leg rec (°)	0.47	0.13	0.81	0.79	0.27	1.00
12	Knee angle begin leg in-sweep (°)	0.35	0.10	0.61	0.30	0.17	1.00
13	Knee angle begin leg out-sweep (°)	0.40	0.11	0.68	0.57	0.13	1.00
14	Maximum hand displacement (m)	0.01	0.00	0.01	0.70	0.15	1.00
15	Head displacement at breathing (m)	0.35	0.10	0.60	0.65	0.34	1.00
16	Trunk angle during breathing (°)	0.01	0.00	0.02	0.63	0.20	1.00
17	Hand disp. end arm in-sweep (m)	0.01	0.00	0.02	3.12	0.68	1.00
18	Elbow angle end arm in-sweep (°)	2.39	0.68	4.10	0.32	0.09	1.00
19	Elbow angle end arm out-sweep (°)	2.54	0.73	4.36	0.15	0.06	1.00
20	Hand disp. end arm out-sweep (m)	0.01	0.00	0.02	0.69	0.33	1.00
21	Hip angle to vertical end leg rec (°)	1.14	0.33	1.96	1.33	0.29	1.00
22	Knee angle end leg rec (°)	1.06	0.30	1.82	1.85	0.73	1.00
23	Knee disp. end leg rec (m)	0.01	0.00	0.02	1.59	0.29	1.00
24	Feet disp. end leg rec (m)	0.01	0.00	0.02	1.78	0.56	1.00
25	Ankle angle begin leg out-sweep (°)	1.93	0.55	3.31	0.39	0.15	1.00
26	Knee angle begin leg out-sweep (°)	2.22	0.63	3.7	1.30	0.31	1.00

	Technique variable	Digitising Error			Inter-trial reliability		
			95% CI				
		SD	Low 95%	High 95%	Typical error %	Shift in mean %	ICC
27	Knee angle begin leg in-sweep (°)	2.08	0.59	3.56	0.94	0.28	1.00
28	Knee disp. begin leg in-sweep (m)	0.01	0.00	0.02	1.94	1.04	1.00
29	Foot disp. begin leg in-sweep (m)	0.01	0.00	0.02	1.34	0.62	1.00
30	Arm phase timing (s)	0.07	0.02	0.11	6.49	1.75	0.88
31	Leg phase timing (s)	0.05	0.02	0.09	5.63	2.57	0.92
32	Average velocity (m/s)	0.05	0.01	0.08	2.94	1.31	0.97
33	Stroke length (m)	0.02	0.01	0.03	1.91	0.71	0.99
34	Stroke rate (m and s)	0.02	0.01	0.04	3.23	1.02	0.97

3.3.4. The Exclusion of Technical Variables for Future Analysis

After applying the exclusion criteria explained in section 3.2.5, fourteen technical variables and swim time (a total of fifteen variables) were identified for inclusion and use in the following chapters of this thesis. The measurements of the accuracy, precision, digitising error and test-retest trial reliability of each variable which were smaller than the worthwhile change are shown in Table 3.8.

Table 3.8: A summary of the technical variables which met the criteria for inclusion. SWC = smallest worthwhile change, SD = standard deviation, CI = confidence intervals, TE = typical error, ICC = intra-class correlation.

Variable	SWC	Accuracy	Precision	Digitising error		Test-retest trial reliability		
				SD	95% CI range	TE	Shift in mean	ICC
Max foot disp. (m)	0.01	0.01	0.00	0.01	0 (0-0.01)	0.01	0.00	0.99
Max hand disp. (m)	0.01	0.01	0.00	0.01	0 (0-0.01)	0.00	0.00	1.00
Trunk angle during breathing (°)	0.97	0.00	0.00	0.35	0.25 (0.10–0.60)	0.34	0.18	1.00

Variable	SWC	Accuracy	Precision	Digitising error		Test-retest trial reliability		
				SD	95% CI range	TE	Shift in mean	ICC
Head disp. at breathing (m)	0.02	0.00	0.00	0.01	0.01 (0-0.02)	0.01	0.00	1.00
Hand disp. arm out-sweep (m) L	0.07	0.03	0.03	0.01	0.01 (0-0.02)	0.01	0.01	1.00
Hand disp. arm out-sweep (m) R	0.08	0.03	0.03	0.01	0.01 (0-0.02)	0.01	0.00	1.00
Knee angle leg rec (°) L	0.04	0.03	0.03	0.01	0.01 (0 – 0.02)	0.01	0.00	1.00
Knee angle leg rec (°) R	0.03	0.03	0.03	0.01	0.01 (0-0.02)	0.01	0.00	1.00
Foot disp. leg insweep (m) L	0.03	0.03	0.03	0.01	0.01 (0-0.02)	0.01	0.00	1.00
Foot disp. leg insweep (m) R	0.03	0.03	0.03	0.01	0.01 (0-0.02)	0.01	0.00	1.00
Leg glide phase timing (s)	0.04	0.04	0.04	0.03	0.02 (0.01–0.04)	0.04	0.02	0.96
Average velocity (m/s)	0.06	0.01, 0.04	0.01, 0.04	0.05	0.03 (0.01–0.06)	0.05	0.02	0.97
Stroke length (m)	0.07	0.01	0.01	0.02	0.01 (0.01–0.03)	0.04	0.01	0.99
Stroke rate (m/s)	0.04	0.01, 0.04	0.01, 0.04	0.02	0.02 (0.01–0.04)	0.04	0.01	0.97

3.4. DISCUSSION

The main aim of this chapter was to investigate and establish a series of technical variables in breaststroke swimming which were valid and could be measured precisely, accurately and reliably using Dartfish software and 2-D video analysis methods. A list of technical measures, classed as errors which could influence the biomechanical performance of swimming, were compiled using the literature. Only those factors which could be measured with confidence were included in the final measurement list and this included fourteen key technical variables and swim time (a total of fifteen variables out of a total of thirty-five) to be used in this thesis. Specifically, coaches must be made aware of this when they are using video analysis as a feedback tool for their athletes. It was stated that to provide high quality data, measurements must be capable of being taken reliably, precisely and with accuracy. This study has shown the final measurements to be capable of measuring the technical variables reliably, accurately and precisely. The validity, accuracy, precision, and reliability of these measurements will now be discussed.

3.4.1. The Validity of Measurements

To ensure that the data collected is valid, it is important that values measure what they are supposed to measure (Atkinson and Nevill, 1998). Large associations were found between angle and distance measurements using Dartfish software and criterion measures. The SEE for distance values ranged from 0-0.02m, with the higher SEE present for out-of-plane measures. For all angle values the SEE was less than 1° when the measure was in-plane, however out-of-plane this increased to over 1°, and over 6° for the front view perspective. These values were similar to related research which also displayed high correlation values for concurrent validity of Dartfish for hip and knee angles with no statistically significant differences in comparison to goniometric measures, $r > 0.95$ (Norris and Olson, 2011); and timing measures in comparison to a recording system, $r = 0.99$ (Mytton et al., 2013). Balsalobre-Fernández et al. (2014) also found similar results in assessing the jump height and time using Kinovea software (Kinovea Association, France, www.kinovea,org) or a jump mat ($r = 1.0$, statistically significant to $p < 0.0001$). Finally, Dingenen et al. (2014) found moderate statistically significant correlations of knee valgus angle and lateral trunk motion when compared with peak external knee movement during a single leg jump; $r = -0.36$ (dominant leg) and $r = -0.32$ (for non-dominant leg) to $p < 0.05$. Although statistically significant, these correlations were only found to be of moderate strength. This may be due to use of a front view camera perspective and the complications of the limb positions during the leg actions. There was no mention of the quantification or minimisation of potential perspective errors in this study and this may be one of the reasons for the differences to the present study due to the use of a correction factor. Although previous authors, including Leblanc et al. (2009); Schnitzler et al. (2009), Alberty et al. (2008); and Alberty et al. (2011) have used 2-D video analysis in swimming research, they have focused on timing and lower limb angle

measures. The present study identified that an additional number of angle, distance and timing measures using 2-D video analysis while swimming are also valid.

3.4.2. The Accuracy and Precision of Measurements

The quantification of errors in any measurement is important so that results of any analysis can be stated with a degree of certainty (Payton and Bartlett, 2007). Findings from this study show that measures taken in-plane of the camera are more accurate and precise than those out-of-plane. The worst measures were those taken out-of-plane and from a front view perspective with more than double the errors arising for both distance and angle measurements. This is an important point to consider as many coaches will often observe from a front-view perspective (especially when coaching using middle lanes). Therefore, these measurements will not be used in this thesis and coaches are recommended to be cautious when observing swimmers from this perspective and to avoid quantitative analysis without corrections using such approaches. Garhammer and Newton (2013) noted the accuracy of 2-D video analysis methods from a sagittal plane recording of a snatch in weightlifting. The authors considered the 2-D software accurate if they were less than 1cm of the horizontal displacement (1cm) or 19cm/s of the maximal vertical velocity. Although different measurements were taken in the current study, the distance values correspond to the requirements imposed by Garhammer and Newton, with distance values (apart from the front view perspective) being within 0.01m for accuracy and precision. Additionally, Alberty and colleagues stated an accuracy of 0.02m in the analysis of stroke phase timings in swimming (Alberty et al., 2008, Alberty et al., 2011). The video data in these studies were recorded at 50 frames per second, however in the present study data were recorded at 25Hz. The use of a frame rate of 25Hz may explain the lack of stroke phase timings which were found to be accurate or precise and the different findings in previous studies. Capturing an image every 0.04s may have resulted in missing the key points between stroke phases which lasted very short durations. With a faster frame rate it could be hypothesised that stroke phases could have been identified more accurately and precisely. This may have then enabled the identification of changes in specific stroke phases, such as an increase in the recovery phases of the arms (Seifert and Chollet, 2005). Although a higher frame rate is preferred, this was not capable on the video system utilised in the present study.

The larger errors in the front view perspective may have been for several reasons; firstly the calibration used was a simple scaling object

(as using 2-D software) and thus any action outside of the calibrated plane (perpendicular to the optical axis of the camera) could be a source of error (Bartlett, 2007); secondly as the object was moving towards the camera, perspective and parallax errors may have been apparent. These occur when there is a difference in length of the limbs which are closer to the camera or at an angle to the photographic plane (Bartlett, 2007). Finally, misplacement of anatomical landmarks or skin movement during the recording may also have caused inaccurate measurement of angles or distance measures. These errors were attempted to be minimised by using multiple calibration scaling measures, the investigator having a thorough knowledge of anatomical landmarks and utilising a side and front view correction factor. Although the scale factor showed a strong, statistically significant association r=0.9997, at $p < 0.05$, with very low variability (SEE = 0.01), large inaccuracies and imprecision were still found for measurements from the frontal camera. Due to the limited literature in this area it was imperative that these factors were considered and quantified in the present thesis. These limitations could be potentially reduced by developing the approaches used in the current study, such as the calibration and scaling method and the use of a more detailed correction factor (the current study only corrected every 0.25m) but further research is required to develop these processes and understand the implications this may have for 2-D video analysis. In the meantime, coaches should take caution in the measurements they make using Dartfish software and 2-D video methods, particularly those taken from a frontal perspective as these produced the greatest errors. It is advised coaches only use measurements which can produce data of a high standard to maximise the information they are receiving.

3.4.3. The Reliability of Measurements

To assess the reliability of data, measurements must be objective and consistent within trials and the study (Atkinson and Nevill, 1998). In the present study this was assessed by test-retest reliability (the variability between trials) and intra-rater digitising error. The digitising error, indicated by the SD and 95% confidence intervals of the measurement error, were deemed to be low for the majority of technical variables. Larger values of standard deviation were found for angle measures taken from a frontview camera perspective. Part of the reason for this may be due to the selected limbs not being exactly in-plane with the camera which, as stated previously, are also the measures with the

greatest inaccuracy in measurement. In terms of the digitising error between trials the average typical error, the random variation between tests, and the average shift in mean between trials had low values for all variables. This indicated a high consistency between trials. Very few studies have utilised these measures in the analysis of reliability. Tan et al. (2010) noted the reliability of measuring the total time of a waterpolo sprint set trial with a typical error of 0.44s. When the decrement in performance time was assessed it yielded a typical error of 0.55s. These values were much greater than the typical error established in the present study. This may be due to the length of the recording being analysed, with the current study analysing smaller time periods. As no other research has analysed the reliability of angle, distance and time measures using 2-D video analysis methods in swimming, this study is the first to determine the reliability of kinematic measures specific to technique performance in a swimming environment.

The consistency amongst trials in the present study was also evidenced by the high ICC values found for each variable of $r > 0.90$. However, the stroke phase timing variables, which made up the total arm or leg stroke timing values had low ICC values (ranging from -0.10 to 0.88) in comparison to the other technique measures. This may have been due to the recording sample rate of the video cameras and the small time gaps between some of these phases. Literature on static and dynamic lower limb angles during functional tasks (Norris and Olson, 2011); body angle measures, distances and timing in weightlifting actions from a sagittal view (Rucci and Tomporowski, 2010); and trunk, thigh and knee angles during curling (Kraemer, 2009) all found similar high values of test-retest reliability of $r = 0.79$ or higher using either Dartfish (Dartfish Ltd, Fribourg, Switzerland) or Silicon Coach (Silicon Coach, Dunedin, New Zealand) 2-D video analysis software. It is also important to note the value of using more than one statistical approach to determine the inter-trial reliability in this study (Atkinson and Neville, 1998). If a single approach was used such as the ICC, all the variables would have been shown to be reliable. However, the high percentage typical error showed that the horizontal alignment at the end of the arm recovery (18.93%) varied between trials and may not be a reliable measurement, as shown in Table 3.7. The use of only one method could have been misleading in the identification of reliable measures. The high reliability in the measurement of angle, distance or timing measurements using Dartfish software, and the applicability of this software for use by coaches, indicate the potential of the use of this method in the monitoring of swimmers during training.

3.4.4. The Smallest Worthwhile Change of Measurements

To ensure that the measures to be included in this thesis produced useful and meaningful data, the varying aspects of error, reliability and validity were compared to the smallest worthwhile change. The smallest worthwhile change is a value used to quantify that measurements are sensitive enough to establish small changes, particularly found in sports performance (Atkinson and Nevill, 1998). This is especially important in swimming, where aspects cannot be easily replicated in the laboratory (Fulton et al., 2009). A number of other studies have utilised a similar method in their use of 2-D video analysis software called the smallest detectable difference (Dingenen et al., 2014, Munro et al., 2012). It was calculated using a 95% confidence interval range and the standard error of measurement. This is deemed as the smallest change in score between tests that can be regarded as statistically significant (Fulton et al., 2009). Although the smallest worthwhile change has been used in research throughout a variety of sports, a limited number of authors, such as Pyne et al. (2006) and Trewin et al. (2004), have utilised the smallest worthwhile change in swimming to assess whether small changes in performance are meaningful. However, this is the first study to have used the smallest worthwhile change for breaststroke 2-D kinematic measurements. The use of the smallest worthwhile change in the present study was demonstrated as being capable of determining whether any changes in the technical measures were small enough to determine a worthwhile change or were simply a result of measurement error.

3.5. CONCLUSION

The present study has successfully achieved its aims by establishing a series of technical measures which are valid and can be reliably, precisely and accurately quantified using 2-D methods and Dartfish software for use in this thesis. These technical measures thus meet the criteria required to support their use in the analysis of swimming technique in the studies reported in subsequent chapters of this thesis.

Chapter 3: Summary

What was already known about this topic?

- Ensuring errors in measurements are minimised is an important part of maximising the accuracy and effectiveness of athlete feedback and the research process.

- Previous studies have shown that angle, distance and time measurements can be accurate, reliable and valid when measured during simple sporting actions.
- Literature reporting the quality of data obtained from 2-D video analysis methods is sparse.

What new information does this chapter provide?

- A series of fourteen swimming technique variables and swim time can be precise, valid and reliable when measured using Dartfish software.
- Perspective and parallax errors can be reduced using correction factors for certain distance measurements.
- Certain technique measurements should be monitored with caution due to large errors.
- Angle and distance measures taken out-of-plane of a camera's optical axis in swimming are not as valid, reliable or precise as those taken in-plane of a camera.

4

An Investigation into the Kinematic Changes in Breaststroke Technique during a High-Intensity Training Set

4.1. INTRODUCTION

The ability to achieve or sustain a maximal swimming velocity is determined by the technique of a swimmer (Lees, 2002, Maglischo, 2003). Swimmers attempt to adopt optimal positioning and orientation of the body and limbs in an effort to reduce the water's resistance to forward motion (drag) and maximise propulsion within the physiological constraints (Toussaint, 2011). The technical skills required to maximise a swimmer's velocity are learned and can become automated through years of training where swimmers may cover up to 16,000m a day (Arellano and Gavilan, 1999, Maglischo, 2003, Sweetenham and Atkinson, 2003). The first swimming technique to be developed was breaststroke, and all the remaining swimming techniques were adapted from it.

Breaststroke is the slowest of the four competitive swimming strokes (Maglischo, 2003) due to the underwater arm and leg recovery actions which cause large amounts of drag (Leblanc et al., 2009). This can cause swimmers to lose considerable momentum during these stroke phases (Maglischo, 2003). This results in breaststroke having large fluctuations in velocity with both large propulsive and decelerating forces occurring within each stroke cycle as swimmers attempt to maintain and maximise their swimming velocity (Seifert and Chollet,

2005, Seifert et al., 2011a). Accordingly, breaststroke has been shown to have the highest active drag and be one of the least economic and most physically demanding swimming strokes (Barbosa et al., 2006, Kolmogorov et al., 1997). As with the other strokes in swimming, it is therefore imperative that stroke technique and mechanics are performed correctly to maximise the propulsion achieved in this stroke (Craig et al., 1988) and minimise the drag. Due to continuous changes in the rules governing breaststroke swimming, it has also undergone a number of changes in styles over the past fifty years (Seifert et al., 2011a). For example, rules allowing the head to go below the surface of the water, other body parts to break the surface of the water, and most recently a single butterfly kick during the underwater phase of the start and turns, all have implications for technique style and drag.

Over the years a number of different breaststroke styles, including; vertical, flat, undulated, and undulated with overwater recovery of the arms, have all been identified and used in competitive swimming (Chollet et al., 2004, Persyn et al., 1992, Tourny et al., 1992). A more in-depth description and breakdown of the breaststroke styles, actions and phases are available in many swimming textbooks (please refer to Maglischo, 2003 or Sweetenham and Atkinson, 2003). After the rule changes in breaststroke technique from the mid 1960's onwards, researchers began attempting to identify the differences amongst these styles (Czabanski and Koszcyc, 1979). Differences amongst the breaststroke styles have been found in terms of leg actions, limb movements, and the relative durations of stroke phases, body segment angles and energy expenditure (Colman et al., 1998, Persyn et al., 1992, Sanders, 1996, Van Tilborgh et al., 1988). Since the identification of these differences, researchers have attempted to ascertain why such differences in style are present. Colman et al. (1998) noted that differences in style could also relate to the swimmers' capabilities. More recently, Persyn et al. (2005) ascertained that the differences in styles may be related to individual morphology of strength and flexibility characteristics. They surmised that swimmers using a flat breaststroke style had high ankle flexion and knee and hip outward rotation whereas swimmers using an undulating style had greater trunk flexibility (Soons et al., 2003). Despite the notion that the flat breaststroke style may be more economical and require less energy expenditure (Vilas-Boas and Santos, 1994), athletes continue to use the undulating style (Cappaert, 1996, Sanders, 1996, Seifert et al., 2011a). From this literature, it appears that no single style is suitable for all swimmers and individual variation must be taken into account in terms of analysis and training prescription. Further research is needed to investigate individual or collective technique factors to aid the monitoring of technique during training.

Over the past forty years, one of the most prominent areas of research in biomechanical analysis of breaststroke technique has been on stroke phase durations, timing or coordination and their link to swimming velocity. Nemessuri and Vaday (1971) were the first researchers to quantify stroke phases and the cyclical activity of the arms and legs. This revealed that stroke phases were associated with fluctuating velocity within the stroke cycle and overall swimming velocity (Bober and Czabanski, 1975, Miyashita, 1971, Tourny et al., 1992). Subsequently a number of studies have been completed investigating the stroke phase durations, arm-leg coordination and intra-cycle velocity patterns of breaststroke (Chollet et al., 2004, D'Acquisto, 1988, Leblanc et al., 2009, Leblanc et al., 2007, Seifert and Chollet, 2005, Strzała et al., 2014, Vilas-Boas, 1996). The arm-leg coordination and intra-cyclic velocity variation were highly associated with the leg glide phase duration, which was found to be the most variable stroke phase, with some swimmers utilising a long glide phase and others a short glide phase (Manley and Atha, 1992). This variation is thought to be due to differences between swimmer's physical properties, technique styles, race distances and paces and even competition levels (Chollet, 2007, Takagi et al., 2004, Tourny et al., 1992). These studies have predominantly compared swimming techniques, genders (Chollet et al., 1996, Seifert and Chollet, 2005), race distances (Strzała et al., 2014, Takagi et al., 2004), swim speed and performance levels (Leblanc et al., 2010) and determined certain distinguishing features. To prevent deceleration during the non-propulsive phase, glide time duration and more efficient arm-leg coordination were shown to be distinguishable features of elite performance swimmers (Seifert et al., 2006, Strzala et al., 2013, Vilas-Boas, 1996). The longer 200m event distance has also been shown to result in a decreased SF, increased SL, increased glide time and more coordinated propulsive movements compared to 50 and 100m breaststroke events (Seifert et al., 2006, Takagi et al., 2004). Finally, in terms of gender, men cover a greater distance per stroke, and are deemed to possess better streamlining and more effective timing between completing the kick and beginning the arm propulsion phase (D'Acquisto and Costill, 1998). Further research is needed to address these factors by investigating technique parameters both between and within swimmers.

One area of continued controversy is the point of highest velocity within a stroke cycle. Previous research has found the point of highest velocity or propulsion to be attained during the leg kick (Capitão et al.,

2006, D'Acquisto and Costill, 1998, D'Acquisto, 1988), some during the arm (Loetz et al., 1988, Manley and Atha, 1992), and some to be similar (Maglischo, 1982, Tourny et al., 1992). This topic was reviewed in further detail by (Seifert et al., 2011a) and it was concluded that the controversy in these measures may be due to differences in methodology or the technical characteristics of the participants.

From the existing literature, biomechanical methods used to investigate breaststroke technique have varied. These include; qualitative (Loetz et al., 1988), 2-D analysis (Leblanc et al., 2009, Seifert and Chollet, 2005), digitisation and 3-D reconstruction (Strzała et al., 2014). Some of the stroke phases were defined from angular positions and others from the intra-cyclic velocity variations of the centre of mass by 3-D analysis or speedometers on the hip joint (Colman et al., 1998, Craig et al., 1988, D'Acquisto and Costill, 1998, Leblanc et al., 2007). Colman and Persyn (1993) noted the issues with such measures due to restrictions in observation as a result of water movement while swimming. Over recent years additional equipment has been developed and utilised alongside video to assess kinematic variables and intra-cyclic velocity variations of the centre of mass (COM), with the intent of providing more applicable information to coaches and athletes including velocity-video systems using tethered swimming and the measurement of instantaneous velocity alongside video feedback on display monitors (Craig et al., 2006, D'Acquisto and Costill, 1998). However, tethered swimming can restrict some movements and potentially provide data which may not replicate how swimmers perform in training or competition. Video continues to be a non-invasive method of investigating swimming technique, however many of the methods described in the existing literature contain complex, expensive equipment which may not be accessible to coaches and athlete support staff. The previous chapter (Chapter 3) established that 2-D video analysis methods could be used to provide similar valuable information to coaches in an efficient and applicable manner. Further research is needed to determine whether these methods can also be used to assess any changes or adaptations in technique during training.

It is the specific technical actions with which a swimmer performs a stroke that determine stroking parameters, efficiency and thus the speed of their swimming. However, the current literature has mainly focused on stroking parameters and race-like distances of 50, 100, and 200m swims only. In addition, it is not apparent how these parameters would change during a race or training-like scenario. Previous literature has identified that swimmers change stroking parameters

(Dekerle et al., 2004, Marinho et al., 2006, Toussaint et al., 2006); alter stroke coordination (Alberty et al., 2005, Tella et al., 2008, Toussaint, 2007) and make specific technical alterations to their limb positioning (Deschodt et al., 1999, Suito et al., 2008) in an attempt to compensate and cope with fatigue. Although these studies highlight the influence of fatigue in swimming, the results are limited to upper body actions or stroking parameter changes during frontcrawl or butterfly swimming during race-like scenarios. Despite the breadth of research pertaining to breaststroke technique, the research investigating breaststroke technical factors and fatigue, or in fact fatigue in any of the four swimming strokes, is scarce. Conceição et al. (2014) analysed fatigue during a 200m breaststroke race pace swim. The focus of the article was on neuromuscular, physiological and biomechanical measures.

However, only the changes in technical stroking parameters were investigated. The sample size was also small, with only nine swimmers taking part. Further research is needed to understand the performance of breaststroke technique during training-like scenarios and any subsequent influences of fatigue on technical actions throughout the entire body.

Understanding the performance of breaststroke technique and influences of fatigue during training-like scenarios is important for several reasons. Firstly, an important goal of training is to optimise performance by allowing swimmers to refine the skills necessary to achieve successful performance (Richmonda et al., 2015). It is during training that athletes learn how to produce patterns of muscle recruitment necessary for optimal technical performance. The interaction between the neural and muscular systems is fundamental to all movement and can result in effective athletic performance (Bonacci et al., 2009). Secondly, skills, such as technical performance, are not thought to be fully learnt until they can be performed with high levels of consistency (McMorris, 2014). Thus continued practice of a skill, through training, helps neuromuscular adaptations which lead to skilled control of movement and improved technical performance (Bonacci et al., 2009, Handford et al., 1997, Jurimae et al., 2007). It is a common belief that to develop expertise in sport requires intensive practice spanning many years (Ericsson et al., 1993, Handford et al., 1997, Williams and Hodges, 2005). Williams and Hodges (2005) noted that before international performance can be reached in sport, an excess of 10,000 hours of practice is necessary.

It is important to appreciate that any adaptations in motor recruitment and coordination as a result of training and practice represent a learning effect. This includes those actions which are practiced incorrectly due to the effect of fatigue. Therefore, dedicating time to improving stroke technique during practice may prevent or reduce the risk of overuse injuries (Richmonda et al., 2015). To achieve effective learning, Ericsson et al. (1993) suggests pitching clearly defined activities at the appropriate level of difficulty, providing effective feedback and the opportunity for repetition, error detection and correction. In order to be able to achieve this, it is important to continue and research further the technical skills and errors which can occur during training.

4.1.1. The Purpose of the Study

It is apparent that further research is needed to provide a deeper understanding as to how a swimmer's technique is altered (or struggled to be maintained) in an attempt to maintain performance during high-intensity training sets. Understanding the individual and common changes in technique due to fatigue is vital to develop and individualise training programmes to minimise the effects of fatigue which are detrimental to swimming performance and maximise effective training time, particularly in breaststroke as it is one of the most demanding of the four competitive swimming strokes. The purpose of this study was to establish whether a series of 2-D technical markers can be used as indicators of acute fatigue during breaststroke swimming under training conditions. The purpose was addressed by:

- Investigating the effects of fatigue on kinematic technical markers during breaststroke swimming in elite national level swimmers.
- Determining whether the changes in technical markers throughout the highintensity set are similar among elite national level breaststroke swimmers.

It was hypothesised that:

- A series of 2-D technical markers will change during a high-intensity, breaststroke set which mimics training conditions as swimmers adapt their technique in an effort to maintain a high performance level.
- Technical markers will change differently between individual swimmers

4.2. METHODS

4.2.1. Participants

The eighteen breaststroke swimmers described in Chapter 3 also participated in this study (see Chapter 3, table 3.5). Prior to the test session, participants provided written informed consent (or participant's parents/guardians if under sixteen years old) approved by the Edinburgh University Ethics Committee; completed the participant information form; the 'pre-activity medical questionnaire'; and the maximal testing statement detailed in Appendix 3 and Chapter 3. Participants were also provided with an information document, detailing the purpose of the study, the test session protocol, and the potential risks and benefits of being involved in the study.

Throughout the study, each participant was asked to continue their normal dietary habits and ensure they were well rested and hydrated prior to their test session to minimise the effects of these confounding variables on their performance.

4.2.2. Experimental Design

This study used a repeated-measures research design, whereby each participant completed one test session, during the early preparation phase of the swimming training season. This design was selected due to the individual differences between swimmers in terms of physiology and technique (Bielec and Makar, 2010, Thomas et al., 2011, Turner et al., 2008). The dependent variables were the fourteen breaststroke technique measures and swim time identified in Chapter 3 (See Appendix 2 for a full description), 25m and 100m lap times, heart rate and rating of perceived exertion (RPE). All these were measured every 25m for each participant, over a series of 100m breaststroke swims which formed the fatigue test set. The independent variable was the number of laps that each participant completed. The participants were made aware that their swimming performance was being analysed but were blinded to the total number of 100m swims they were required to complete by not specifying the number of 100m swims to be completed. This was done to avoid the tendency to pace throughout the set.

Data were collected over a period of two weeks, in an attempt to prevent history and maturation effects on the validity of the data (Thomas et al., 2011). The collection of data during the early preparation phase meant each participant's training volume and intensity were lower than other training phases and helped to enable swimmers to attend the test session in a non-fatigued state. In addition, participants were asked not to train on the morning of their test session and to reduce the volume and intensity of their training 24 hours prior to the test session.

The test session lasted one hour, between 9am and 1pm, to control for diurnal effects on each swimmer's performance (Kline et al., 2007) and to coincide with a time block in which the pool was available.

Each participant's test session was allocated two weeks in advance to minimise impact on their training schedule. Each participant attended their test session individually to prevent the data collected being influenced by the presence of other participants and were asked to refrain from talking about the session until the end of the data collection phase to ensure those participating later in the study remained blinded to the protocol. One swimmer withdrew from the study due to contracting a chest infection immediately prior to their test session (identified by the medical questionnaire).

4.2.3. Data Collection Protocol

The data were collected in the same pool and conditions as described in section 3.2.4. Participants wore brief swimming trunks or costume and were asked not to wear additional 'training' trunks to remove this confounding variable, ensuring swimmers wore the same kit, and reducing any additional drag effects the trunks may offer. Once swimmers had changed into their swimsuits, their height and body mass were recorded using a stadiometer (Seca 220, Seca, Birmingham, UK) and weighing scales (Seca 803, Seca, Birmingham, UK), respectively. Each participant was then marked with black waterproof actors' paint applied with a 3cm circular sponge at nine anatomical landmarks and joint centres, on both sides of the body, as described in section 3.2.4 (Bartlett, 2007). These markings were used to aid the identification of the anatomical landmarks and body midline during video analysis.

Each participant completed the thirty minute warm-up as described in section 3.2.4 to ensure they were prepared for the test session and to reduce the risk of injury (Balilionis et al., 2012). After a five minute period of passive rest, participants completed 5x25m breaststroke at 100m race pace on three minutes, from a push start, to establish baseline measures of each athlete's technical performance at the test session. As described in section 3.2.4, the correct swim pace was checked using the split times against each individual's 100m split; participants swam back to the start end of the pool completing both active and passive recovery rest methods; and each participant was assessed to ensure they were working within 5% of their personal best.

After completing the 5x25m swims, participants were given a further five minute period of passive rest and then asked to complete a protocol designed to mimic a high-intensity training scenario and induce a stressed state. This consisted of a series of maximum effort 100m breaststroke swims, from a push-start, on a swim-rest interval of each participant's 100m personal best plus thirty seconds.

Swimmers continued repeating 100m swims until the time to complete the 100m became greater than 125% of the swimmer's personal best time or voluntary cessation, representing a fatigued state. For each of the 100m swims, the participant was encouraged to complete the swim as close to target pace as possible each time. This protocol was effective in eliciting a stressedinduced state resulting in changes in many technique variables while allowing for individual differences in performance as established by Thow (2010). It was also designed to mimic standard training sets swimmers often undertake (Maglischo, 2003). The test session was concluded with a swim down, during which the participant was free to swim for as long as they deemed necessary. Pilot work carried out prior to the study established the total session time, preparation requirements and the manageability of this data collection protocol.

4.2.4. Data Collection Methods

Each swimmer's technical performance during each 25m and 100m swim was video recorded using the 2-D video set up and calibration methods described in section 3.2.1 and modified from Thow (2010). Pilot work was carried out to ensure that the position of each camera was sufficient to record two stroke cycles at the centre of the swim lane and maximise the accuracy and reliability of the technical variables measured from this video, as introduced and established in Chapter 3.

Throughout the research, the intensity of the swim set and fatigued state of each participant was established using heart rate, an RPE Borg scale (Wallace et al., 2009) and performance swim times. The heart rate was recorded using a heart rate monitor (Polar Accurex Plus, Polar Electro Oy, Kempele, Finland), worn at all times throughout the test session, which recorded each participant's heart rate every 5 seconds throughout the test session (Turner et al., 2008). An RPE Borg scale was also used to indicate the intensity each swimmer felt they were working during the set. This was noted at the end of each 25m and 100m swim. The performance times were recorded manually using a stopwatch (Finis 3x-100m, Finis, California, USA). Times were noted for each 100m swim as well as splits taken every 25m.

4.2.5. Data Analysis Methods

Each outcome measure was analysed every 25m. As described in Chapter 3, each technical variable was analysed using Dartfish Pro suite motion analysis software (version 6.0, Dartfish, Fribourg

Switzerland) over two stroke cycles. As per the results from Chapter 3, only those variables which could be measured reliably, precisely and accurately were included in the present study (fourteen technique measures and swim time, fifteen dependent variables in total). A description of how each technical variable is analysed using Dartfish is available in Appendix 2. The calibration method and corrective scale factor used in this study are also described in Chapter 3.

Fatigue was determined by the deterioration in performance swim times (Maglischo, 2003). The recorded 100m and 25m split times were compared to each individual's personal best 100m breaststroke time and the first 25m of the 100m swims (Thow, 2010). The participants were defined as 'fatigued' if their 100m swim time deteriorated more than 20% of their personal best time. This cut off percentage was used as it signifies deterioration greater than that experienced within a 100m – 200m competition breaststroke event at elite level, and thus would stress the swimmers to more than race performance (Thow, 2010). These data were established from the swimming results of international competitions (available at: http://www.swimming.org/britishswimming). The deterioration in swim time was established by calculating the percentage increase in 100m swim times relative to each individual's personal best 100m breaststroke time (for example, a swimmer with a personal best time of 60s, who swims a time of 66s, would have a time increase of 10%). The percentage increase in 25m split times throughout each 100m swim was also calculated relative to the average of the 5x25m swims (Thow, 2010). This aided the understanding of changes in swimming performance between and within each 100m swim.

The peak heart rate from each 100m swim was identified from the heart rate recordings and, along with the RPE Borg scale value, indicated the intensity of the swim set (Achten and Jeukendrup, 2003, Turner et al., 2008). Although no single method has been shown to indicate exercise intensity effectively, these methods were chosen due to their ease of application and history of association with exercise intensity prescription and description (Achten and Jeukendrup, 2003, Borresen and Lambert, 2009).

4.2.6. Statistical Analysis

All statistical analysis was completed using the Statistical Package for Social Sciences,

SPSS (version 19.0, IBM UK Limited, Portsmouth, UK). Descriptive statistics, including the mean and standard deviations, were calculated for all the data in Microsoft Office Excel 2010 software.

The data were pre-analysed to assess the number of stroke cycles that could be used to represent each swimmers 'normal' swim technique. Previous literature has varied between using one to three stroke cycles per length as an indicator of individual swimmers' technique. To validate the number of stroke cycles per length to be used to indicate each swimmers' technique in the present study, a repeated measure ANOVA, with the number of stroke cycles as the factor, was used to evaluate whether the number of stroke cycles used as an indication of normal stroke technique differed between an average of two, four, six, eight and ten stroke cycles. The data were assessed for normality using a Shapiro-Wilks test and Sphericity was assessed using Mauchly's test. If the assumption of sphericity was violated a Greenhouse Geisser correction was used (Field, 2009). This was repeated for each technical variable. The data were found to be normally distributed and no statistically significant differences were identified ($p > 0.05$). Therefore, the mean of two stroke cycles was used for each lap.

4.2.6.1. Comparison of technical variables

The effects of fatigue on kinematic technical markers were analysed using a paired t-test to compare the first 25m of the first 100m to each individual's last 25m of the last 100m swam for each technical variable, with an alpha level of $p < 0.05$ accepted as statistically significant (Thow, 2010). Due to the within subject repeated measures nature of the study and use of only one independent variable, a paired t-test was used (Field, 2009). The data were assessed for normality using a Shapiro-Wilks test. Any data that were not normally distributed were assessed non-parametrically using a Wilcoxon signed rank test (Field, 2009).

In addition to the paired t-tests, a 95% confidence interval (CI) of the true mean was quantified for each variable to also assess whether the change in technique was statistically significant or not (Hopkins, 2000). The upper and lower CI boundaries were presented to indicate the range in which the true value of the change in the variable falls 95% of the time. The CI were quantified using the formula: $CI = d \pm t\ (SD) / (\sqrt{n})$ (Portney and Watkins, 2000, Thomas et al., 2011), where d is the sum of the mean difference between the first and last 25m for each individual, t is the t-score for a 95% confidence interval with a degrees of freedom of sixteen, a value of 2.12, SD is the standard deviation of the differences and n is the number of subjects (Portney and Watkins, 2000). The degrees of freedom was sixteen due to the sample size of seventeen participants. Due to the small sample size (n = seventeen), a t-value was used within the CI formula (Portney and Watkins, 2000).

To establish the magnitude of the change in each technical variable between the nonfatigued and fatigued state for each variable, an effect size was calculated (Sullivan and Feinn, 2012). The general effect size formula is given as: effect size = $M_1 - M_2/SD$ (Thomas et al., 2011). In the literature (Coe, 2002), it is unclear which dataset should be subtracted from the other when no control group is used thus it is important to quote which order the calculations are performed. In this study, M_1 is the mean of the two stroke cycles from the first 25m swim of the fatigue set, for each participant. M_2 is the mean of the two stroke cycles from the last 25m of the fatigue set, for each participant. Due to discrepancies over which standard deviation to use when calculating the effect size, a pooled standard deviation of both data sets was used, see Equation 4 below.

$$S_p = \sqrt{\frac{(n_1 - 1)S_1^2 + (n_2 - 1)S_2^2}{n_1 + n_2 - 2}} \qquad \text{Equation. 4}$$

Where S_1 is the variance of M_1 (the first dataset), S_2 is the variance of M_2 (the second dataset), and n_1 and n_2 is the number in each data set respectively. These values were interpreted according to Hopkins (2002) where 0 is trivial; 0.2 is small; 0.6 is moderate; 1.2 is large; 4.0 is very large; and infinite is perfect. Effect sizes were deemed suitable as they normalise for differences in magnitude and within participant variability of scores and indicate the relative importance and measurability of change for each outcome measure and individual swimmer (Coe, 2002, Thow, 2010). It is also deemed vital to report an estimate of meaningfulness in all tests of significance (Sullivan and Feinn, 2012).

To assess which variables relate to the change in swim speed as a result of fatigue, a backwards multiple linear regression was used. In using the backwards method, all the predictor variables are placed into a model which calculates the contribution each variable has made to the model (Field, 2009). This is assessed by comparing the significance value of a t-test for each predictor variable to the removal criteria (Field, 2009). If the variable is not making a statistically significant contribution and is the weakest predictor variable (meaning it has the lowest partial correlation), it meets the removal criteria (based on the software, SPSS) and it is removed (Field, 2009). This method is preferred over the forward method, particularly for exploratory research, as there is less chance of making a type II error (Field, 2009). Only variables that were found to be statistically significant, had a confidence interval that did not cross zero and an effect size of greater than 0.2 (or small) were candidates for inclusion in the regression. The difference between the first and last lap of the 100m swims was calculated for each variable. This difference was the value entered into the regression with the average change in swim time entered as the dependent variable and the remaining technical variables entered as the independent variables. To assess the multicollinearity of the variables the correlation matrix was scanned of all predictor variables to assess if any correlated very highly. In addition, the variation inflation factor (VIF) and the tolerance values of each variable were assessed to determine if a predictor variable had a strong relationship with any other predictor variable.

4.2.6.2. Comparison of each individual's technique

To determine whether changes in the technical markers were consistent among elite breaststroke, two factors were considered:

- The lap number at which the technique was first established outside the 'normal range' (indicated by the lap number).
- The number (%) of laps which were outside the established 'normal range' were identified for each individual in each technical variable.

To indicate the change in technique, the range in which the true value of each individual's 'measure for that variable' or 'swim style technique' was calculated using 95% confidence intervals of the 5x25m swim results. The confidence intervals were calculated using the formula: CI = X ± (t x SD) (Field, 2009, Portney and Watkins, 2000). X is the mean of the 5x25m swims, t is the t-score for a 95% confidence interval with a degrees of freedom of nine, a value of 2.262, SD is the standard deviation of the 5x25m swims, for each individual. As the CIs were completed for each individual, the number of subjects was always one. The degrees of freedom are based on the ten measurements from the two stroke cycles of each 5x25m swim. The standard deviation was used instead of the standard error as the variance of interest was the individual's swim style over the 5x25, not the group or population (Portney and Watkins, 2000). A t-value was used instead of a Z-score due to the small sample size (n = ten, the lap numbers) (Portney and Watkins, 2000). The 95% confidence interval of the 5x25m swims was used to indicate the individuals 'non-fatigued' swim technique style. From this, the results of each 25m lap completed during the fatigue set was compared to the 95% confidence interval range to determine if they were within or out with this range. If the value was outside this range (upper or lower) this established that the technique variable was not within the established 'normal range', that is, the effect of fatigue was statistically significant.

4.2.6.3. State of fatigue and intensity of exercise

The individual deterioration in performance swim time, was presented descriptively as percentages, means and standard deviations. These included the total time decrement from the first to last 100m, for each individual, and the percentage decrement of the final 100m time to each participant's personal best time. To indicate the intensity each athlete was working at, the peak HR and the RPE value from each 100m swim were determined for each swimmer. The mean and standard deviation for the group were determined using Microsoft Office Excel 2010 software.

4.3. RESULTS

4.3.1. Comparison of Technical Variables

From the fourteen technique variables and swim time (fifteen dependent variables in total) identified in Chapter 3, all data were normally distributed ($p > 0.05$) and eleven variables were shown to have statistically significant changes ($p < 0.05$) throughout the high-intensity swim set (the first to final lap), as shown in Table 4.1. Those variables which were statistically significant are in bold. The mean change between the first and final lap was much larger for the leg glide duration and four stroking parameters, with the leg glide duration, SF and swim time increasing, and the average velocity and SL decreasing. To also interpret how meaningful these changes were, the effect sizes were calculated. These values ranged from a trivial effect size of 0.03 (left knee displacement at leg recovery) to a very large effect size 4.04 (25m swim time). The largest effect sizes were again found for the leg glide duration, and the four stroking parameters, indicating their change was very meaningful (See Table 4.1).

Table 4.1: A summary of the technique changes. The mean change, standard deviation, significance value and absolute effect size of each technical variable. SD = standard deviation, P = alpha significance value, L = left, R = right, * = statistically significant

Technical variable	Mean change ± SD	P value	Absolute effect size
Foot displacement	-0.01 ± 0.04	p = -0.14	0.22
Hand displacement	-0.04 ± 0.02	p < 0.001*	0.70
Head displacement at breathing	0.02 ± 0.05	p =0.40	0.17
Trunk angle during breathing	1.84 ± 2.16	p < 0.001*	0.39

Technical variable	Mean change ± SD	P value	Absolute effect size
Hand displacement arm out-sweep L	-0.02 ± 0.03	p = 0.19	0.05
Hand displacement arm out-sweep R	-0.01 ± 0.04	p = 0.03 *	0.04
Knee displacement leg rec L	0.01 ± 0.03	p = 0.54	0.03
Knee displacement leg rec R	0.02 ± 0.03	p = 0.01 *	0.13
Foot displacement leg in-sweep L	-0.04 ± 0.03	p < 0.001 *	0.21
Foot displacement leg in-sweep R	-0.03 ± 0.04	p = 0.01 *	0.17
Leg glide phase	0.31 ± 0.23	p < 0.001 *	1.30
Average velocity	-0.2 ± 0.19	p < 0.001 *	2.22
Stroke length	-0.49 ± 0.15	p < 0.001 *	1.86
Stroke rate	0.42 ± 0.21	p < 0.001 *	0.58
25m swim time	5.64 ± 1.35	p < 0.001 *	4.04

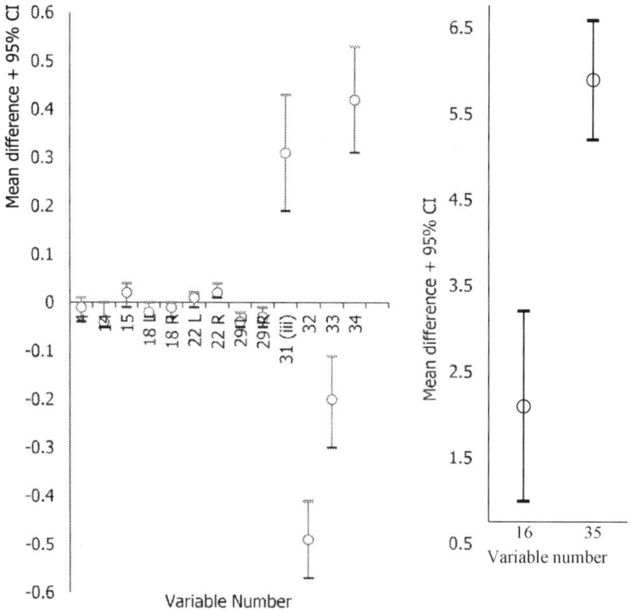

Figure 4.1: The mean difference and 95% confidence limits of each technical variable between the first and last lap. Each variable number corresponds to the technique variable specified in Appendix 2, L = left, R = right, CI = confidence limits.

The statistically significant change in technique was also indicated and emphasised by the fact that those technical variables which statistically significantly changed also had a 95% confidence interval which did not cross zero, another indication that the changes are statistically significant (See Figure 4.1) (Hopkins, 2000). The large confidence limits for the trunk angle, leg glide time, average velocity, SL, SF and 25m swim time shown in Figure 4.1 show the large changes between the first and last lap for these particular variables.

Table 4.2: A summary of the multiple regression analysis. The multiple regression output related to change in 25m swim time pre and post a high-intensity test. B = unstandardized beta value; SE B = standard error of unstandardized beta value; β = standardised beta; r = the correlation; R^2 = correlation variability; VIF = variation inflation factor. * $p < 0.05$, 1, 2, 3 and 4 relate to each model level.

Model	**B**	**SE B**	**β**	**Tole-rance**	**VIF**			
Time (Constant)	2.76	1.46	-					
Hand depth	7.70	13.7	0.14	0.77	1.30			
Trunk angle	-0.80	0.15	0.12	0.78	1.28			
Leg glide time	-4.66	3.96	0.79	0.11	9.56			
Stroke length	-3.42	2.42	0.48	0.41	2.41			
Stroke rate	7.36	5.10	1.16	0.07	13.6			
Average velocity	-1.96	2.69	0.22	0.52	1.94	Model 1 summary		
						r	R^2	Change R^2
						0.73	0.53	0.53
2 – Time (Constant)	2.44	1.26	-					
Hand depth	5.70	12.6	0.10	0.85	1.18			
Leg glide time	-4.89	3.79	0.83	0.11	9.42			
Stroke length	-3.90	2.14	0.54	0.49	2.03			
Stroke rate	8.05	4.73	1.27	0.08	12.6			
Average velocity	-1.55	2.47	0.18	0.57	1.75	Model 2 summary		
						r	R^2	Change R^2
						0.72	0.51	-0.01

3 – Time (Constant)	2.15	1.05	-					
Leg glide time	-4.72	3.64	-0.80	0.11	9.33			
Stroke length	-4.16	1.99	-0.58	0.53	1.88			
Stroke rate	8.11	4.57	1.28	0.08	12.6			
Average velocity	-1.43	2.37	-0.16	0.58	1.73	Model 3 summary		
						r	R^2	Change R^2
						0.71	0.51	-0.01
4 – Time (Constant)	2.37	0.97	-					
Leg glide time	-5.68	3.20	-0.97	0.13	7.56			
Stroke length	-4.80	1.64	-0.67*	0.74	1.35			
Stroke rate	9.69	3.66	1.53*	0.12	8.47	Model 4 summary		
						r	R^2	Change R^2
						0.70	0.49	-0.02

Table 4.2 presents the initial test model, and subsequent models used to assess the relationship of these variables to the outcome variable, the 25m swim time, and their ability to predict this. The correlation matrix did not show any predictor variables with a correlation of $r > 0.8$, indicating they did not show a strong relationship with each other.

The last model, number 4, statistically significantly improved the ability to predict the outcome measure ($p < 0.028$) (Field, 2009). In this model, no predictor variables had a VIF value of greater than ten or a tolerance value of less than 0.1, also indicating that the predictor variables did not show a strong relationship with each other (Field, 2009). The relationship between Model 4 and the predictor variables (leg glide time, SL and SR) was found to be moderate as indicated by the correlation coefficient value $r=0.70$, however this was not statistically significant ($p = 0.56$). The predictor variables within this model also accounted for 49% of the variability of the outcome value. The beta values in Table 4.2 indicate the relationship between the 25m swim time and each of the predictor variables (Field, 2009). The negative values of the leg glide time and SL indicate a negative relationship with the swim time, so that if either of these variables decreased, so too would the swim time. The SF however, showed a positive relationship. Both the SF and SL showed statistically significant t-test values ($p = 0.012$, 0.020, respectively), demonstrating that both these variables were making a statistically significant and similar contribution to the model. The standardised beta value of SF however, is much higher than SL, indicating it has more importance in the model (Field, 2009).

4.3.2. Comparison of Each Individual's Technique Changes

Due to the individual nature of technical actions, each swimmer's technical measures were assessed against a 95% norm based on their own performance. This was very effective at highlighting the differences between individuals in the performance of the technique parameters, as well as the individual responses which occurred throughout the high-intensity swim set within each individual. Each individual had a different 95% confidence range to indicate their technique norm. Each individual also showed values outside this range at different laps throughout the swim set and for different variables. The results from one swimmer are provided in Figure 4.2, using three technical variables. Figure 4.2 highlights that even inside each individual's 'normal range' there were variations between each lap and also that each technical variable demonstrated changes outside of the established 'normal range' at different stages of the set. Only three variables are represented below and the remaining technical variables also demonstrated different changes throughout the set. In addition, the changes in each variable were different amongst the sample of swimmers used in the present study emphasising the individual variations in technical performance.

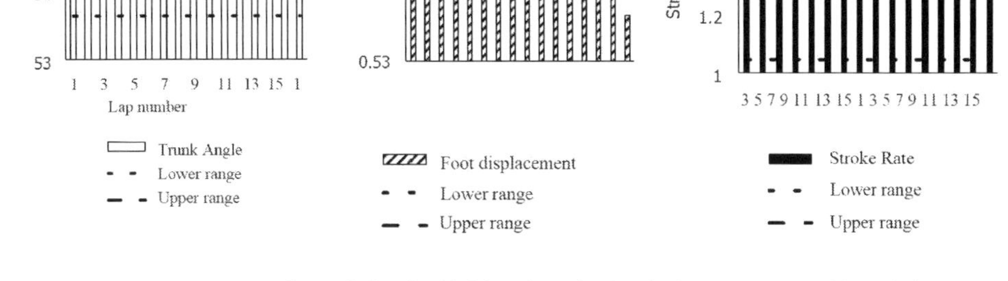

Figure 4.2: Examples of the individual technical changes outside of the measured 95% CI norm of Swimmer 4 for three technical variables: Trunk angle during breathing, maximum foot displacement, and stroke rate. The lower range represents the lower 95%CI and the upper range represents the upper 95% CI.

Although the changes in technique were individual, some variables highlighted commonalities within this group of swimmers. The swim parameters such as average velocity, SL, SF and 25m swim time were shown to change within the first 100m swim (laps 1-4) amongst all the swimmers. The direction of change of these variables was also the same for all the swimmers. The remaining technical variables displayed changes outside the 95% norm during the second-third 100m or onwards (laps 5-12) and either increased or decreased depending on the swimmer, again indicating that individual responses were evident. The mean number of laps for which all of the swimmer's technique remained outside of their normal 95% range was also identified in Table 4.3. This differed depending on the variable, although again the 4 stroking parameters tended to show a higher number of laps outside of their 95% norm.

Table 4.3: A summary of the individual changes in technique. SD = standard deviation, disp. = displacement, L = left, R = right, * indicates statistically significant.

Technical variable	Mean Lap number changes began at (SD)	Mean duration technique changes lasted (lap number) (SD)	Direction of change in technique
Foot displacement	7.29 (5.01)	5.47 (8.69)	Mixed
Hand displacement	5.35 (4.47)	9.24 (5.71) *	Mixed
Head disp. at breathing	4.12 (2.78)	10.00 (7.71)	Mixed
Trunk angle during breathing	6.29 (3.82)	8.82 (7.65) *	Mixed
Hand disp. arm out-sweep L	5.59 (4.69)	6.24 (9.10)	Mixed
Hand disp. arm out-sweep R	9.06 (6.68)	5.76 (3.96) *	Mixed
Knee disp. leg rec L	9.00 (4.87)	3.76 (4.07)	Mixed
Knee disp. leg rec R	8.76 (6.41)	4.65 (3.32) *	Mixed
Foot disp. leg in-sweep L	7.59 (3.41)	6.35 (7.71) *	Mixed
Foot disp. leg in-sweep R	8.00 (5.48)	5.06 (4.67) *	Mixed
Leg glide phase	5.12 (4.55)	9.24 (5.12) *	Increases
Average velocity	2.65 (2.09)	11.35 (7.93) *	Decreases

Stroke length	3.47 (2.94)	7.88 (3.67) *	Decreases
Stroke rate	2.76 (1.71)	10.18 (3.97) *	Increases
25m swim time	1.59 (0.51)	13.71 (7.23) *	Increases

4.3.3. State of Fatigue and Intensity of Exercise

Table 4.4 shows the variables measured to represent the state of fatigue of each swimmer and the intensity of the swim set. All swimmers showed a change in swim time ranging from 4.51s to 21.82s from their first 100m swim to their last. In terms of deterioration, relative to each individual's personal best, the group showed an average change of 27.75 (9.11%). This is well within the cut-off percentage to represent a fatigued state greater than that shown in competition. Participants all showed a heart rate of greater than 168bpm, indicating that the swim set was high-intensity.

Table 4.4: The swim times and physiological results (heart rate and RPE). SD = standard deviation, s = seconds, BPM = beats per minute, RPE = rate of perceived exertion.

	Change in time (s)	**Time decrement (%)**	**Peak HR (BPM)**	**RPE**
Group mean (SD)	10.55 (4.92)	27.75 (9.11)	182.75 (8.72)	19 (0.87)
Range	4.51 to 21.82s	16.71 to 43.50	168 to 201	17 to 20

This was verified by the RPE Borg scale which showed an average response throughout the set of seventeen or higher (very hard intensity), with the majority of swimmers stating they felt the swim set was extremely hard (level nineteen). During Swimmer number 9's set, the heart rate monitor unfortunately failed to record and this heart rate data was removed, However, it can still be inferred that this swimmer was swimming at a high-intensity due to an RPE Borg scale of nineteen (very hard intensity) and a deterioration of 15.53s between the first and last 100m swim time, a percentage of 32.28%.

4.4. DISCUSSION

This study successfully distinguished that a single high-intensity training set can affect the performance of certain kinematic technical markers and that these changes can be different between individual swimmers. The technical markers were identified through literature as errors in the stroke which could theoretically influence the swim speed through the effect of biomechanical factors such as drag or propulsion. The highintensity training set highlighted changes in eleven technique parameters amongst all participants with meaningfulness ranging from trivial to very large. This included;

- The right hand displacement at the end of hand out-sweep
- The right knee displacement at the end of leg recovery
- The left and right foot displacement at the beginning of leg in-sweep
- The trunk angle during breathing
- Stroke rate
- The maximum hand displacement
- Leg glide duration
- Stroke length
- Average velocity
- 25m swim time

The dependent variables utilised in the present study were identified as being accurate, valid and reliable measures in Chapter 3. The largest difference in the present study was the 25m swim time, which showed changes ranging from 3.75-8.77s between the first and last 25m lap. These changes had a meaningful effect size of 4.04. This implies that the changes in 25m swim time were large meaningful changes. In addition to this, changes in the 100m swim time ranged from 16-40% of each individual's personal best time. The changes in swim time suggest that the high-intensity swim set had a large impact on the swimmers' ability to sustain their maximal swimming velocity and were therefore, as per the definition of fatigue used in this thesis, fatigued.

The technical variables which portrayed the next largest changes related to stroking parameters such as SF, SL and swim velocity. Research has identified that stroking parameters are key technical factors which change when swimmers experience fatigue (Marinho et al., 2006, Stirn et al., 2011). The link between stroking parameters and reduction in swim velocity and time was evident by the statistically significant relationship of SF and SL to the change in 25m swim time. Similar correlations were found between SF and SL as those found in Thompson et al. (2000), with slightly lower values found in the present study. The different values may be a result of the different measurement methods, such as Thompson et al. (2000) analysing only the finishing time whereas changes during the set were investigated in the present study. This enabled assessment of the relationship between SF and SL and whether this relationship changes throughout a session.

Similarly, a greater speed at the beginning of the test session was associated with a higher SF and a shorter SL, as in previous literature (Conceição et al., 2014, Strzała et al., 2014, Takagi et al., 2004). As the test session progressed, the SF increased and the SL decreased, again similarly to previous literature (Aujouannet et al., 2006, Conceição et al., 2014, Hellard et al., 2008, Marinho et al., 2006). Simultaneously, as the swimming speed decreased, different SF and SL combinations were observed as the swimmers attempted to adapt and maintain an optimal swimming speed as a result of fatigue (Conceição et al., 2014). The decrease in swimming speed and SL, and the increase in SF in the last lap, suggested decreased capacity to generate propulsion and an attempt to compensate for decreased propulsion in each stroke by increasing the SF (Conceição et al., 2014, Seifert and Chollet, 2005). This implies that in addition to changes in technique factors during training, there may also be other changes which could result in the decrease in swim time and performance, as suggested by the literature (Ament and Verkerke, 2009). Further research is required to investigate other changes during training as a result of fatigue in other areas of sport science and the implications these may have for technical performance.

The current literature analysing changes with fatigue have focused on competition or race-like scenarios. In race scenarios, changes in stroking characteristics are considered a strategy used to address changes in constraints, including fatigue (Hellard et al., 2008, Suito et al., 2008). Suito et al. (2008) identified that when fatigued, individuals compensate by changing certain arm actions and as a result arm pull velocities decrease during the pull phase in front-crawl swimming but they did not identify any acute technical changes. Although breaststroke is considerably different from front-crawl, the similar changes in stroking characteristics suggests that swimmers attempt to make adaptations regardless of the technique they are using in an effort to maintain an optimal swimming velocity. It would also imply that the compensations observed during race scenarios are similar to those apparent during high-intensity training. This is the first study to investigate and identify the effects of fatigue on SL and SF in breaststroke swimming during a training-like situation. Further research is needed to assess the effects of fatigue on SL and SF in the remaining three strokes of swimming during training-like situations using similar 2-D video methods.

In addition to the stroking parameters, the breaststroke technique variables which were shown to change with fatigue in the present study included distance measures of the limbs at set phases of the stroke, and leg glide duration. Only leg glide duration showed a statistically significant and meaningful change in performance during the test session. This is in accordance with previous breaststroke literature which has shown the glide duration to be the most variable stroke phase, with high positive correlations found between glide time and the duration of arm propulsion and the velocity decrease during arm and leg recovery (Manley and Atha, 1992, Seifert et al., 2011b, Tourny et al., 1992), and a key indicator of skill level amongst swimmers (Manley and Atha, 1992, Sanders, 1996, Tourny et al., 1992). Currently there is no other literature in swimming which has looked at such technical factors in a training environment. Further research is required to ascertain which specific technical changes have the largest effect on swimming performance in all four swimming techniques, both in race and training environments.

The protocol used within this study was specifically designed to mimic a high-intensity training session often experienced by swimmers which would result in a fatigued state (Glaister, 2005, Maglischo, 2003, Thow, 2010). The fatigued state of each individual was objectively determined and confirmed by increases in performance times throughout the test session. The increase in performance time was comparative to that experienced during a competition environment (www.swimming.org/britishswimming). Four swimmers ceased swimming voluntarily before achieving their 20% cut-off (see section 4.2.5). Despite this, these swimmers had a mean peak heart rate of more than 170bpm and indicated, using the RPE Borg scale, that they found the activity to be more than level seventeen (very hard intensity). These factors indicate that the swimmers were still in a fatigued state, despite not achieving the cut-off mark. The reason these individual's may have had to cease prior to the predetermined 20% cut-off may be related to individual differences in technical style. As stated by Vilas-Boas (1996) and Barbosa et al. (2008), individual differences in technical style could impose a higher or lower energy expenditure per stroke cycle. This suggests that individuals with a certain technical style could have expended a higher amount of energy and thus were unable to maintain their work intensity for as long. This may explain why certain individuals were able to continue swimming for a longer duration than others. However, without direct measurements this is only a suggestion. Additional reasons for the individual differences in the level of fatigue experienced by the athlete, include: fitness capacity, motivation, or even nutrition (Enoka and Duchateau, 2008). The maximal heart rate values and exhaustive perceived exertion (RPE) scores recorded throughout the fatigue test indicated the high-intensity of the protocol (Achten and Jeukendrup, 2003, Borresen and Lambert, 2009, Wallace et al., 2009). These were similar to those reported in previous studies (Glaister, 2005, Marinho et al., 2006, Thow, 2010).

Although fatigue during training is an important part of the training process in terms of athlete adaptation, it is a concern that through continued practice during high-intensity training sessions, athletes may adapt or alter their technique in an effort to maintain performance. As a result, muscular adaptations may result in incorrect muscle development, imbalances and increased injury risk (Kluemper et al., 2006). The identification of technique patterns due to fatigue before they become automatic is imperative to maintain peak performance and prevent muscular imbalances that reduce performance or increase risk of injury. Understanding and managing fatigue during training is therefore an important part of the training process, a process controlled and directed by the coach. Currently, coaches understanding of fatigue during training and their management of this in swimming is unknown. Based on the findings of this study, the changes in these technique variables during high-intensity training can be monitored using Dartfish software and underwater video equipment. To understand the role of 2-D video analysis in monitoring fatigue during training, it is important to investigate coaches' current perceptions of fatigue and if or how coaches currently monitor it during training.

4.5. CONCLUSION

This study distinguished changes in kinematic technical variables during a high-intensity training set using 2-D video analysis methods. These findings highlighted that swimmers adapt and alter their technique during high-intensity training sessions in an effort to maintain a high performance during training. Leg glide time, SF, SL, average velocity and 25m swim time displayed changes which were common amongst all the swimmers. Differences were identified between individuals in terms of when changes in technique outside of the established 'normal range' began to occur throughout the training set. Due to the individual differences and importance of monitoring technique during training, it is important to determine whether these changes in technique can be identified by coaches during training. This information could then be used by coaches and other sport science personnel to guide training programmes, reduce the risk of injury and improve swimming performance.

Chapter 4: Summary

What was already known about this topic?

- There is a wealth of literature on breaststroke technique and its performance during race-like scenarios.
- Research pertaining to monitoring technique and technique adaptations during training in swimming is scarce.
- Monitoring training is important to manage athlete progression effectively, promote the development of technical performance and prevent injury.
- There is a lack of research using 2-D video analysis methods.

What new information does this chapter provide?

- A high-intensity set is sufficient to induce changes in the performance of technical variables.
- Stroke length, stroke rate and leg glide time are related to swim performance time.
- Leg glide time, stroke rate, stroke length, average velocity and 25m swim time displayed changes which were common amongst all the swimmers.
- Individual differences existed in the variation of technique variables outside the established 'normal range'.
- Both technical measures and individual variations were measurable using 2-D video analysis methods.

5

Fatigue During Training: The Perception of Swimming Coaches

5.1. INTRODUCTION

In order to attain the optimal swimming performance in training and competition, swimmers must first develop the necessary skills and prepare themselves physically and mentally. This is achieved through training, globally represented as *'the physical, technical, intellectual and psychological preparation of an athlete through physical and mental training'* (Smith, 2003; p. 1104). To achieve continued adaptations and development, coaches progressively push athletes to a higher level than previously tolerated, and as close as possible to their genetic limit, by increasing the training load (determined by the intensity, duration and frequency of training) using the principles of training (Smith, 2003). The performance capacity and adaptations experienced as a result can be both positive and negative, and are determined by a number of factors, including: the training load and the recovery process determined by the coach; and the athlete themselves (Smith, 2003, Stone et al., 2007). Understanding and managing this is therefore a vital component of the training process managed by coaches.

Due to the stress of an increased or increasing training load (overload), athletes will inevitably experience fatigue (defined as *'the inability to sustain maximal swimming velocity'*, Alberty et al. (2009; p. 638)) following a single intense training session, intense training period or competition (Robson-Ansley et al., 2009). The process of training overload can result in acute fatigue leading to an improvement in performance (supercompensation) when the training load and recovery process are balanced correctly by the coach (Meeusen et al., 2013). If the training intensity continues, and the coach does not balance the training load and recovery effectively, athletes can undergo negative responses and accompanying decrements in performance (Robson-Ansley et al., 2009). The borderline between optimal and reduced performance is subtle and if not carefully monitored by the coach, and left to continue, it can result in two syndromes known as 'overreaching' or 'overtraining' (Bell and Ingle, 2013, Enoka and Duchateau, 2008). These are both defined as '*an accumulation of training and/or non-training stress resulting in a decrement in performance capacity with or without related signs and symptoms of overtraining in which restoration of performance capacity may take a certain duration*' (Halson and Jeukendrup, 2004 p. 969). These two conditions are differentiated within this definition by either short or long-term decrements in performance and the duration it may take to restore performance capacity (several days to weeks and several weeks to months), respectively (Halson and Jeukendrup, 2004, Meeusen et al., 2013). Both overreaching and overtraining are considered to have a multifactorial aetiology and it must be noted that exercise (training) may not necessarily be the sole causative factor of the syndrome (Meeusen et al., 2013). Although these definitions have potential issues and controversies, they are currently the most accurate, frequently used description of the conditions (Halson and Jeukendrup, 2004, Meeusen et al., 2013). Therefore, preparing for sports performance is not a simple process and requires training to be carefully planned and executed by coaches. This requires coaches to have a thorough understanding of this topic and its application.

Training overload which can result in overreaching or overtraining is perceived by many as a process or continuum (See Figure 5.1). Although long-term consequences of fatigue are seen as detrimental to performance, short-term or functional overreaching is seen as a normal part of training and vital to increase the training load and promote adaptation and performance (Bell and Ingle, 2013). The perceptions and knowledge of coaches in swimming regarding this topic is currently unknown and has not been investigated.

According to Halson and Jeukendrup (2004), if athletes undergo periods of highintensity training without sufficient or appropriate recovery, it may result in symptoms of fatigue and decreased performance. This is a particular issue in swimming which is renowned for its large training load and frequent, multiple high-volume training sessions per day (Richmonda et al., 2015). In swimming training the traditional periodization includes four periods, which can often be repeated throughout the year, designated as follows: general endurance; specific endurance, competitive period and a taper period (Maglischo, 2003). A high volume of workload is often used throughout the periodisation periods in swimming to prepare swimmers for the various disciplines in swimming (Arroyo-Toledo et al., 2013). Coaches need to ensure athletes can get the full potential from training sessions. Therefore, the management of short-term fatigue and recovery during training are essential components in this process of an athlete's development.

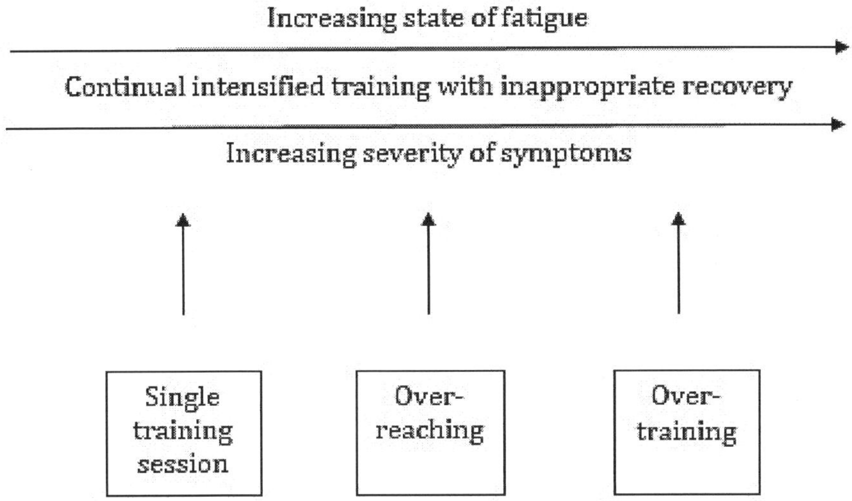

Figure 5.1: The overtraining continuum (Halson and Jeukendrup, 2004).

The general aim of coaching is to develop athletes' performances and prepare them for competition (Dorgo, 2009). This is achieved by the design and implementation of training during which the coach decides what type of practice the swimmer engages in (Smith, 2003). According to Nash et al. (2011), training sessions represent the numerous and wide-ranging skills of a coach and are the point at which coaches bring all aspects of practice and performance together. Part of this role includes the on-going monitoring and management of a swimmer's performance during training to ensure athletes are training and recovering efficiently. There are a range of methods, including physiological, psychological, and biomechanical approaches, which can be used to monitor load and fatigue during training (Smith, 2003, Taylor et al., 2012). Although recently there has been an increase in research on improving coaching interventions and the training process (Debanne and Fontayne, 2009), research pertaining to coaches' knowledge, perceptions and practices regarding monitoring fatigue in training are scarce.

Coaches' knowledge is a key factor affecting coaches' decision making, and thus the way athletes' train and prepare for competition (Dorgo, 2009). This knowledge base is continually developing and adapting as rules change and new methods and developments are established. It also requires coaches to draw on knowledge from several sport disciplines, including the sport sciences, in order to plan and address expected and unexpected issues through improvisation (Kerwin and Irwin, 2008). For coaches to effectively manage and monitor fatigue during training, they require an understanding and knowledge of fatigue, recovery, the principles of training, and other sport science disciplines. To date, no research has investigated coaches' knowledge of fatigue in swimming, nor in any sport for that matter. In addition, although several studies have investigated coaches' knowledge content regarding technique and performance fundamentals amongst expert coaches (Thompson et al., 2009), analysis of coaches' content knowledge in swimming has not been investigated. As current and prospective coaches develop their knowledge and also impart this knowledge to other individuals, including athletes and other coaches, it is of equal importance that the quality of education and information they receive is high (Ozdoğan and Ozcelik, 2011). To date, no research has been conducted on the knowledge of coaches on fatigue and its implications during the training process in swimming.

There are a range of psycho-physiological and performance markers available for use in determining the state of fatigue or recovery of an athlete (Ament and Verkerke, 2009, Enoka and Duchateau, 2008). Although several methods (performance tests, psychological tests and biochemical and immune markers) are currently used by sport scientists in an attempt to monitor the state of fatigue or recovery, none yet meet all the criteria to make their use generally accepted for application during regular training and competition by coaches (Ament and Verkerke, 2009, Meeusen et al., 2013). Anecdotal evidence suggests that high performance sport staff monitor fatigue using a range of markers and use this to monitor their athletes preparation for competition and training (Taylor et al., 2012). Only two other studies have investigated the methods used to monitor fatigue during training. Robson-Ansley et al. (2009) noted during their review of fatigue management in Olympic athletes that often measures, such as training load, psychological mood states and perceived stress, are used to assess fatigue, despite other physiological and biochemical assessments being known, as they can be practically applied to athletes daily in a training environment. Taylor et al. (2012) investigated the type of training monitoring systems currently used to quantify training load (rate of perceived exertion, external workloads, heart rate trimps and heart rate variability, GPS data, and workload measurement devices) and monitor fatigue/ recovery (self-report questionnaires, performance tests, hormonal profiling, and tracking performance). Interestingly Taylor et al. (2012) identified that a high percentage of respondents (70%) focused on monitoring fatigue and recovery during training by mainly using self-report questionnaires (84%) and practical maximal tests (61%). The methods used to achieve this were based mainly on experiential knowledge rather than methods used in scientific publications (Taylor et al., 2012). The respondents ranged from sport scientists, coaches and other individuals only involved at elite level performance, over a wide range of sports in New Zealand; however, this only included a sample size of fifty respondents, even fewer of these being coaches and less than ten individuals representing competitive swimming. Although informative, the majority of research has investigated long-term signs and symptoms of fatigue present in the athlete and very little research has investigated the practices and processes of coaches themselves in the monitoring of fatigue and recovery during training. Since periods of fatigue will compromise performance and training, whether short- or long-term, its management is an integral component of the coaching process and athlete development (Robson-Ansley et al., 2009).

Ensuring that training time is effective yet athletes are recovering (and coping with the stresses of training) is difficult in swimming due to the high and demanding, yet sometimes unnecessary, training load. Despite the role of the coach in training design and implementation, and the importance of fatigue monitoring for performance development and training management, little is known about coaches' understanding of fatigue, nor how they currently manage fatigue during training. The high percentage (29%) of British swimmers developing overtraining syndrome indicate a potential issue in the capacity of swimmers, at varying performance levels, to manage the training load resulting from a training session and recover adequately (Matos et al., 2011). Unfortunately, the focus of research on overtraining can only inform the situation once athletes have already experienced fatigue to an extent which it has been detrimental to their development and performance. Further research is needed to ensure current coaching practices and courses are effective as well as to enable athletes to be able to cope with the high physical and mental demands of training in swimming.

5.1.1. The Purpose of the Study

The purpose of this study was to examine coaches' current perceptions about fatigue and their methods of monitoring it during a training session in competitive swimming. To achieve this, this study aims to:

- Explore coaches' knowledge of fatigue concepts, including the causes, effects and additional factors that can influence fatigue.
- Investigate the types of methods and equipment coaches' use, and the frequency of their use, to monitor fatigue during training.
- Examine the processes coaches use to prevent and manage fatigue during training.

5.2. METHODS

5.2.1. Participants

The participants (n=100) in this study were competitive swimming coaches. Coaches were included in the study if they were currently coaching competitive swimming in the UK, and involved in coaching a squad of competitive swimmers at National age-group performance level or higher. National age-group performance level was defined as swimmers who are currently competing in National level competitions. The coaches who completed the questionnaire were recruited from all over Britain. No personal information was required unless the participant was interested in participating in a follow-up study or receiving the results of the study. Prior to completing the questionnaire, participants were informed of the research purpose and provided informed consent. Upon reading the initial information sheet and continuing to complete the questionnaire they were providing their consent to use the information they provided. The study was approved by the Edinburgh University Ethics Committee.

5.2.2. Questionnaire

A five-section questionnaire was developed using the Bristol Online Survey (BOS, Bristol, UK). The questionnaire was adapted from a coaching survey completed by Nash and Sproule (2012), and based on published scientific literature surrounding fatigue monitoring, training and coaching in swimming (Ament and Verkerke, 2009, Enoka and Duchateau, 2008, Maglischo, 2003). In addition, personal communication with coaches about their current practices and the researcher's experience in competitive swimming provided further basis for construction of the questionnaire. Permission to adapt the survey was obtained from the author of Nash and Sproule (2012) and adaptations were made in the content of questions to make them specific to monitoring fatigue in training. The questionnaire was separated into five sections: Section one was designed to collect demographic information; Sections two-four were designed to investigate coaches' current understanding of the topic of fatigue, the methods coaches employ to monitor fatigue during a training session and the processes used to manage and prevent fatigue impacting their athletes training performance; Section five allowed the participants to provide any additional information. Data were requested in a variety of formats throughout the questionnaire, including open questions, closed questions, and 7-point Likert-scales. A copy of the questionnaire can be located in Appendix 7.

The questions sought to obtain data regarding participants' personal opinions and current practices; there was no correct or incorrect response. Open questions enabled the participants to express opinions and expand on their answers (Portney and Watkins, 2000). As the purpose of the study was to investigate coaches' opinions and practices, it was thought a combination of open and closed questions was necessary to maximise the response rate and obtain as much information on this topic as possible (Thomas et al., 2011). Likert scales were also used in this study due to their easy-to-read presentation and ability to produce a highly reliable scale (Preston and Colman, 2000). According to Preston and Colman (2000), a 7-point Likert scale is the preferred scale than five or less to produce reliable and valid results. Three Likert scales were used in the current study where respondents were asked to rate their familiarity with the mechanisms and effects of fatigue (1 = not at all familiar; 7 = extremely familiar) and the importance they deemed certain additional factors, which can influence fatigue, had on an athletes' ability to maintain peak swim performance during training (1 = not at all important; 7 = extremely important).

5.2.3. Pilot Work

Prior to the study, the questionnaire was piloted among university staff familiar with fatigue and competitive swimming coaches who were not eligible for the study (n=8). This was used to test the comprehension of the questions and the presentation of the questionnaire, determine the time required to complete the questionnaire, and assess the manageability of the data collection and analysis process (Portney and Watkins, 2000). Appropriate adjustments were made to the questionnaire based on pilot responses.

5.2.4. Data Collection Procedure

The questionnaire was distributed online, in association with The University of Edinburgh and the Bristol Online Survey. An online questionnaire was used to enable an efficient method of collating large amounts of data on beliefs, attitudes and practices over a short period of time (Gratton and Jones, 2010, Thomas et al., 2011). The survey data were collected over a period of five weeks during the month of September. The questionnaire was distributed at this time as it is coincides with the early training phase of the competitive swimming season, and it was thought coaches would have more time available to complete the questionnaire due to fewer competitions during this time.

Three main sources were used:

1. Email contact details provided by the SportScotland Institute of Sport;
2. British swimming club websites with coaches' contact details available;
3. Social network sites, including Twitter and Facebook.

Those coaches deemed eligible, and whose contact details were obtained through the provision of email details or swimming club websites, were contacted electronically whereby the purpose of the survey was explained and a link to the online survey was provided. The social network sites were used in an attempt to promote the survey and maximise its distribution to coaches around Britain. A link to the survey was also provided on these sites. Reminder emails were sent to the participants one and two weeks prior to the final deadline. Based on information from the Institute of Sport and British swimming club websites, a total of 374 questionnaires were dispatched to eligible coaches, with a total 165 (44%) responses. However, 65 questionnaires were omitted as they were incomplete, giving a final response rate of 100 (26.7%) questionnaires. Participants were asked not to discuss their results with other coaches or individuals while completing the questionnaire to prevent their answers being influenced by the perceptions or beliefs of others.

5.2.5. Data and Statistical Analysis

All statistical analysis was completed using SPSS (version 19.0, IBM UK Limited, Portsmouth, UK). Descriptive statistics, including the mean and standard deviations, were calculated for all the data in Microsoft Office Excel 2010 software.

The participants' demographic data, as well as the responses in Sections 2 to 4, are presented descriptively as means, frequencies, ranges and standard deviations. As in Nash and Sproule (2012), this provided the most appropriate method of analysing these data. Attitudinal data were collected in the form of a 7-point Likert scale in Sections 2, 3 and 4. A Chi-Square test of association was used to analyse these data and compare responses to determine if there was any relationship between the key variables (Field, 2009, Portney and Watkins, 2000). Other methods of association are available to analyse Likert scale data, however there is little information regarding their use in the literature (Portney and Watkins, 2000). A Chi-square test is a comparison to assess the degree of association between two attributes (Portney and Watkins, 2000). This method is based on comparing the frequencies observed in each category to the frequencies that would be expected if the null hypothesis of no association were true (Field, 2009). In the present study this involved comparing the coaches' coaching qualifications and years coaching to Section 2, the causes, effects and additional considerations of fatigue; and Section 3, the equipment used by coaches and frequency of its use. To ensure the assumptions of a Chi-square test were met, each item only contributed once to each contingency table and analysis was only carried out on contingency tables which showed expected frequencies over 1 (Field, 2009). A limitation of the Chi-square test is that it does not apply well to small samples, especially in a 2x2 contingency table. However, it is easy to apply and is applicable to many problems (Field, 2009, Portney and Watkins, 2000).

The strength of an association found in the Chi-square test was assessed using Cramer's V. This method is deemed acceptable as it provides a method of analysis when the contingency tables are asymmetrical and it is able to achieve a maximum value of 1, regardless of the size or dimensions of the contingency table, which is not available using other statistical methods (Field, 2009, Howell, 2012, Portney and Watkins, 2000).

The participants were also given the opportunity throughout the questionnaire to provide additional information to answers of each question and any additional comments in Section 5. However, as this study was exploratory and due to the shortness of the responses in the survey, the additional comments were used only to add richness to the statistical information gathered.

5.3. RESULTS

5.3.1. Section 1: Demographic and Sport-related Data

Statistical analysis of the survey reveals the following summary of the participating coaches' demographic data:

- 70% males and 30% females, aged between 16-over 65 years old, participated in this study. There were more participating male coaches than female coaches, a ratio of just over 2:1. This is in accordance with previous research analysing coaching (Nash, 2008).
- The highest rate of respondents were from England (47%), followed by Scotland (39%), Wales (8%) and Northern Ireland (6%).
- The coaching qualification level (UK based) ranged from level 1 to level 4, with the largest representation of coaches currently at swim coaching level 3 (38%). The coaches were involved with training swimmers of varying performance levels, from age-group to elite Olympic level and even masters' level. The swimmers of the coaches involved were predominantly aged between 12-20 years.
- The total time in coaching ranged from 1 to more than 25 years, with the majority of coaches having been involved in coaching competitive swimming for 11-15 years.
- Coaches spent between 2-16 training sessions a week (training sessions 7.4 ± 3.1) and 2.5 to 32 hours a week (training hours 13.6 ± 6.80 hours) coaching their swimmers in the pool.

For more information regarding the coaches' demographic details see Appendix 8. There appeared to be a trend between the qualification level of the coach and: the duration of coaching; the performance level of the swimmer; and whether the coach saw their work as full or part-time. The higher the qualification held by the coach, the longer the duration of coaching, the higher the performance level of the swimmer and the more their role was classed as full-time.

5.3.2. Section 2: Knowledge of Fatigue

Figures 5.2 to 5.4 show the percentage of responses by coaches who stated they were extremely familiar with the subsequent causes and effects of fatigue, and perceived the subsequent additional factors of fatigue to be extremely important. This was further broken down into the coaches' qualification levels. The mechanism which had the highest percentage of responses by coaches of 'extremely familiar' was a 'psychological decrease in motivation, interest or enthusiasm' (25%). The mechanism which had the lowest percentage of responses by coaches of 'extremely familiar' and the highest percentage of responses by coaches of 'not at all familiar' was a 'protective mechanism of the body' (10%). A minimum of 17% of coaches with level 3 qualifications and 33% of coaches with level 4 qualifications stated they believed they were 'extremely familiar' with every mechanism of fatigue listed in the survey compared to only 1-10% of coaches with a level 1 or 2 qualification. This, as shown by Figure 5.2, indicates a trend for coaches with a level 3 or 4 coaching qualification to remark they were 'extremely familiar' with a greater range of mechanisms of fatigue than their fellow coaches with lower qualifications.

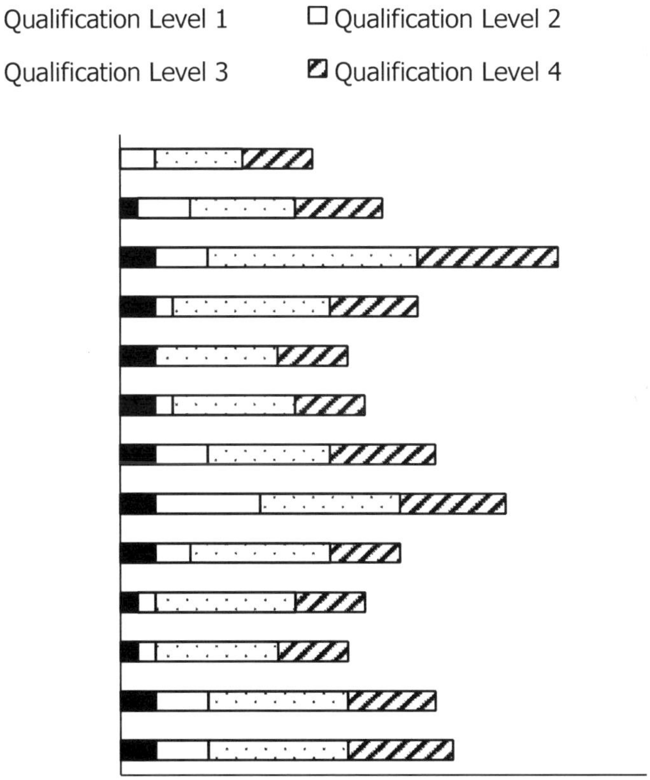

Figure 5.2: The percentage of responses by coaches, and their qualification level, 'extremely familiar' with the mechanisms of fatigue. This figure shows the percentage of coaches and their qualification level who responded they were extremely familiar with the mechanisms of fatigue stated above. RPE = rate of perceived exertion; ATP = Adenosine triphosphate; PCr = Phosphocreatine.

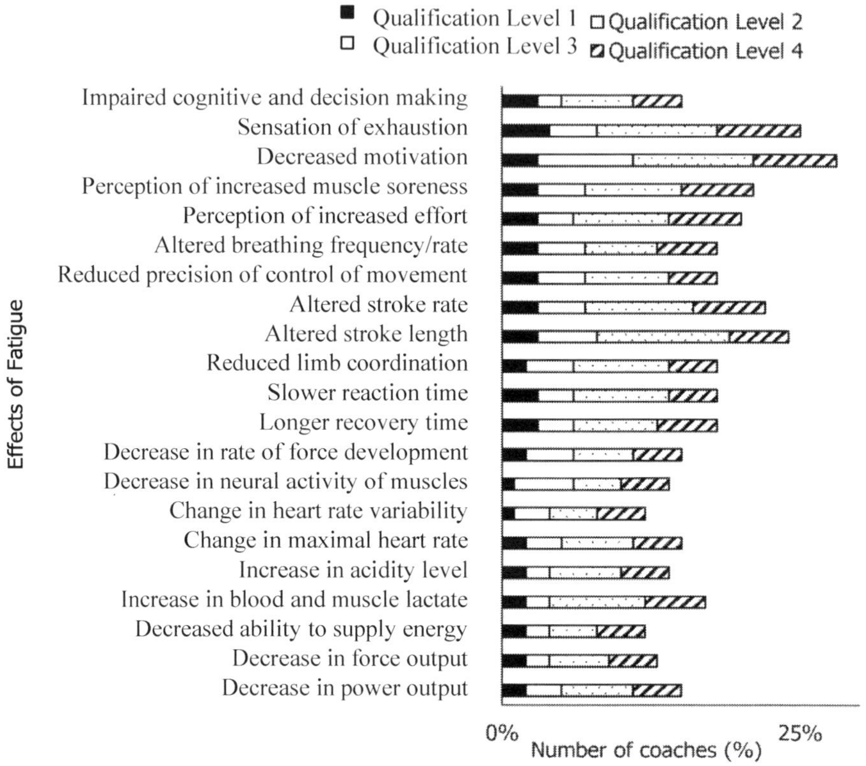

Figure 5.3: The percentage of responses by coaches, and their qualification level, 'extremely familiar' with the effects of fatigue. This figure shows the percentage of coaches who responded they were 'extremely familiar' with the effects of fatigue stated above.

In terms of the effects of fatigue, the highest percentage of responses by coaches of 'extremely familiar' was a 'decrease in motivation, interest and/or enthusiasm' (32%). The effect of fatigue which had the highest percentage of responses by coaches of 'not at all familiar' was a 'decrease in neural activity' (7%). Although it appears that coaches believed they were familiar with the majority of effects of fatigue, as shown in Figure 5.3, it appears more coaches were familiar with physiological effects of fatigue, such as a decreased power output or muscle lactate, than biomechanical and psychological effects. Out of the biomechanical parameters, changes in SL was the effect which the highest percentage of coaches stated they were 'extremely familiar' with. A minimum of 33% of coaches with level 4 qualifications stated they were 'extremely familiar' with every effect of fatigue listed in the survey compared to 7-11% for coaches with qualifications ranging from 1-3. Again, as shown by Figure 5.3, coaches with a level 4 qualification tended to state to be 'extremely familiar' with a larger range of the effects of fatigue than coaches with lower qualification levels. This also indicated that coaches with level 1-3 qualifications stated they had a larger familiarity with the effects of fatigue than the mechanisms of fatigue.

In terms of additional factors believed to influence fatigue, the highest percentage of responses by coaches of 'extremely important' was 'hydration level' (68%). The factor which had the highest percentage of responses by coaches of 'not at all important' was 'gender' (16%). A minimum of 17% of coaches with a level 4 qualification stated they thought each additional factor was 'extremely important' compared to 0-5% for coaches with qualifications ranging from 1-3. As shown by Figure 5.4, coaches with a level 4 qualification tended to perceive a larger range of additional factors 'extremely important' than coaches with lower qualification levels. Similarly, those coaches who had been coaching for a longer period of time appeared to place a higher importance on a larger number of factors believed to influence fatigue.

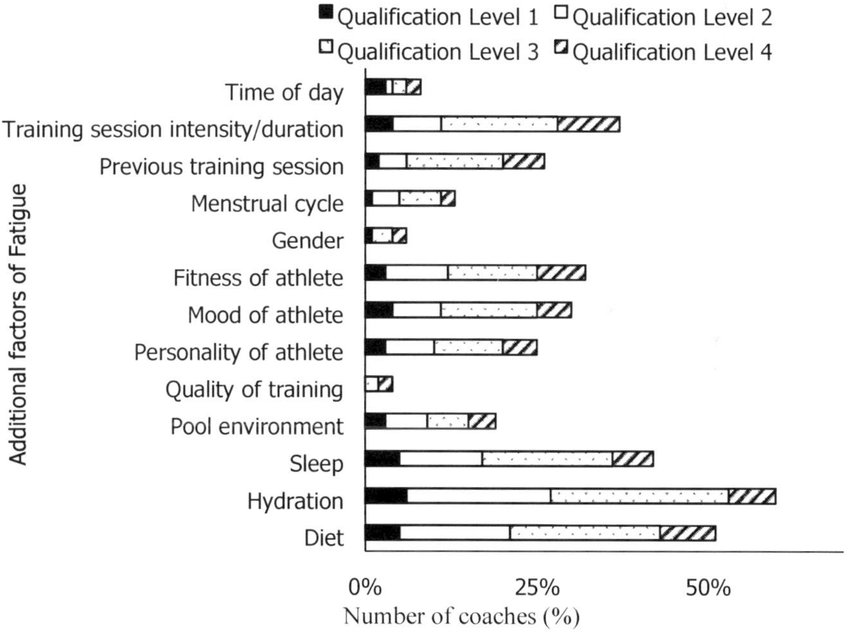

Figure 5.4: The percentage of responses by coaches, and their qualification level, stating the additional factors were 'extremely important' in influencing fatigue. This figure shows the percentage of coaches who responded they perceived the additional factors to be extremely important in influencing fatigue as stated above.

The Likert scale in each sub-section of the knowledge of fatigue was analysed using the Chi-square test and compared to the coaches' qualification levels and years coaching to see if there was a statistically significant association between these factors. Table 5.1 shows the results of the Chi-square and Cramer's V tests which were found to be significant in Section 2, the knowledge of fatigue ($p < 0.05$). The statistically significant associations were found with coaches of a higher qualification level (level 3 or 4), and those coaches who had been coaching for a longer duration (in years), and their knowledge of certain mechanisms or additional factors of fatigue. Although the association was statistically significant, the strength of the association was only small medium, as indicated by values of 0.27 to 0.32.

Table 5.1: The statistically significant associations (Chi-square) and relationships (Cramer's V) of Section 2: Knowledge of fatigue and coaching qualification/ years coaching.

Fatigue causes			
Years Coaching	**Mechanism**	**Chi-Square**	**Cramer's V**
	Reduced energy stores in the short-term	X^2 (36) = 54.5	0.31
	Challenges in the functions of the immediate energy system	X^2 (36) = 51.7	0.29
Coaching Qualification	**Mechanism**	**Chi-Square**	**Cramer's V**
	Challenges in the functions of the high-intensity energy system	X^2 (24) = 37.9	0.31
	Challenges in the functions of the long duration energy system	X^2 (24) = 40.5	0.32
	A psychological increase in RPE	X^2 (24) = 42.1	0.32
Additional fatigue aspects			
Coaching Qualification	**Additional fatigue aspect**	**Chi-Square**	**Cramer's V**
	The quality of training facilities	X^2 (24) = 40.8	0.32
	Fitness of the athlete.	X^2 (16) = 28.6	0.27

The main method by which coaches stated they obtained this knowledge was experience (90%), followed closely by coach education courses (72%). The method least believed to provide knowledge was academic background or personal sporting experience (9%).

This was found to be regardless of the coaching qualification or years of coaching.

5.3.3. Section 3: Monitoring Fatigue

Out of the respondents, 79% percent of coaches stated they monitored the state of fatigue of their athletes during a training session. The most predominantly used form of equipment stated by these 79 coaches to monitor fatigue was the stopwatch (93 %), followed closely by visual observation (91%) and mood questionnaires (86%). A total of 72% and 73% of coaches stated they used heart rate monitors or above-water cameras. A total of 47% of coaches stated they used under-water cameras. Less than 16% of coaches stated they used lactate or blood glucose analysers during training as shown in Figure 5.5.

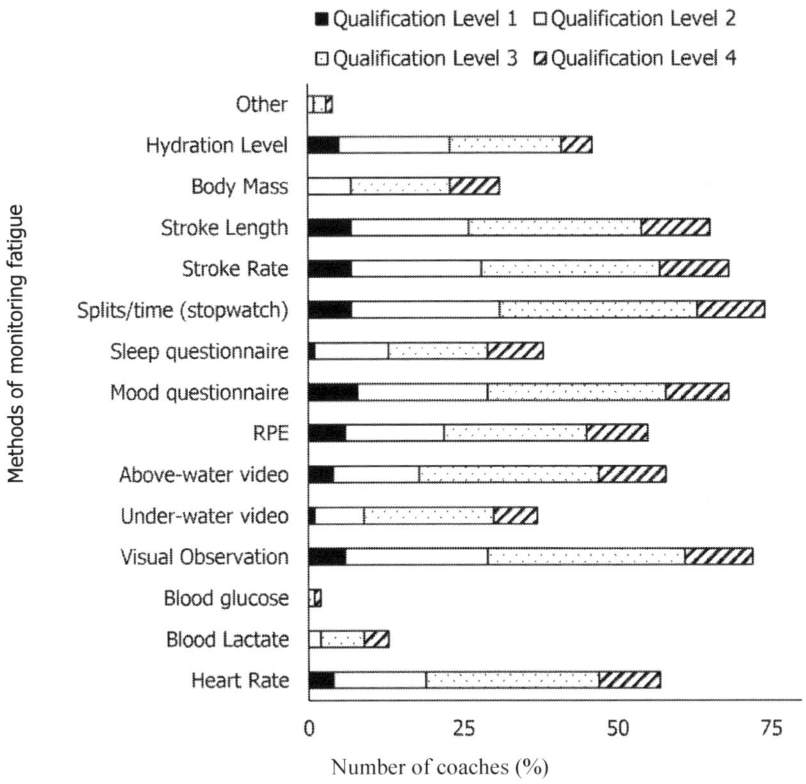

Figure 5.5: The methods used by coaches to monitor fatigue. This figure shows the percentage of coaches who responded they were using each method to monitor fatigue, and their qualification level. RPE = rate of perceived exertion.

Although a high percentage of coaches used swim times and splits as a tool to monitor fatigue only 76% (57 coaches) stated they used this every training session to monitor fatigue. The method coaches stated they used every training session to monitor fatigue was their personal visual observation of technique (98.7%). The methods of assessing fatigue and their use were also analysed using the Chi-square test and compared to the coaches' coaching qualifications and years coaching to see if there was an association amongst these factors. The statistically significant results are shown in Table 5.2.

Table 5.2: The statistically significant associations (Chi-square) and relationships (Cramer's V) of Section 3: Monitoring fatigue and coaching qualification/ years coaching.

Methods of analysis			
Years Coaching	**Mechanism**	**Chi-Square**	**Cramer's V**
	Blood lactate levels Yes	X^2 (16) = 18.53	0.43
	Under-water video Yes	X^2 (16) = 14.50	0.38
Coaching Qualification	**Mechanism**	**Chi-Square**	**Cramer's V**
	Under-water video Yes	X^2 (4) = 13.08	0.36
	Above-water video Yes	X^2 (4) = 16.98	0.41
	Above-water video No	X^2 (4) = 9.51	0.31
	Sleep quantity and quality Yes	X^2 (4) = 10.66	0.33
	Body mass Yes	X^2 (4) = 14.53	0.38
	Hydration level No	X^2 (4) = 9.65	0.31

Although only certain associations were found to be statistically significant between coaches' experience and certain pieces of equipment, there was a trend that the higher the coaching qualification, the more pieces of equipment which appeared to be used. There were some items of equipment that were not used at all by any coach with a qualification level ranging from 1-3, whereas a minimum of 33% of coaches with a level 4 qualification stated they used all of the items of equipment listed in the questionnaire, as shown in Figure 5.5. As the years of coaching increased, coaches stated they used a greater range of methods to monitor fatigue during training. In addition, coaches with a higher qualification appeared to use the equipment more often than those coaches with a lower qualification. Some equipment seemed to be utilised regardless of the coaching experience, such as visual observation, split times and the athlete's mood. Finally, the use of some pieces of equipment showed great variation amongst all levels of coaching qualifications, such as hydration level.

In terms of technical parameters, SL was stated to be utilised by coaches almost every session, however the monitoring of SF using stopwatches or visual observation ranged from every session to almost once a month. In terms of the methods linked to biomechanical technique analysis, the method stated to be most utilised, besides visual observation, was above-water video. Although this was more used by coaches with level

3 and 4 qualifications, the use varied from once a week to once a month with coaches with a lower qualification stating they used this piece of equipment only once a month. The use of under-water video followed a similar pattern, however its use was predominantly by coaches with a level 3 or 4 qualification, as shown in Figure 5.6. Underwater video was only utilised on a monthly basis, even by coaches with higher qualification levels.

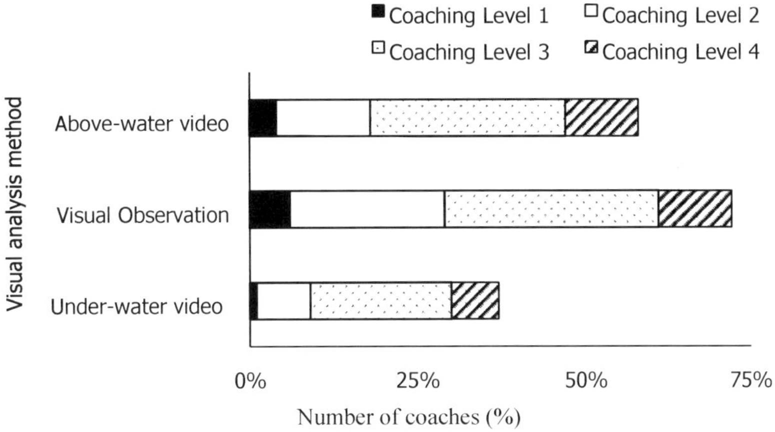

Figure 5.6: The visual analysis methods used by coaches and their qualification level. This figure shows the number of coaches, and their respective qualification levels, using above-, below-water video and visual observation to monitor fatigue.

5.3.4. Section 4: Management of Fatigue

The study found that 98% of coaches stated they made changes to a session plan to make the training set less intense and enable their swimmers to cope with the training session. The changes were focused around the individual (93%) and the main factor changed consisted of the rest duration of the set (92%), which was regardless of the coaching experience, as shown in Figures 5.7 and 5.8.

A high percentage of coaches (over 94%) stated they perceived the effect and management of fatigue during training to be important. As the qualification level decreased, the percentage of coaches monitoring fatigue was also shown to decrease.

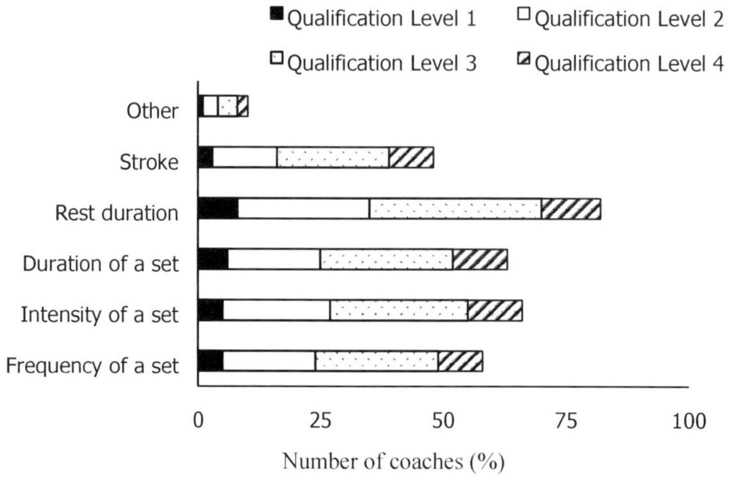

Figure 5.7: The factors coaches use to adapt training sessions. This figure shows the various training factors that coaches use to modify their training if they find their athletes experiencing fatigue.

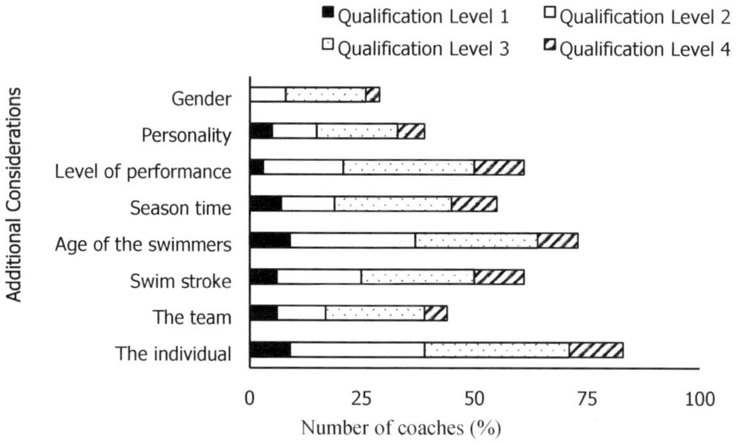

Figure 5.8: The aspects coaches consider when making training adaptations. This figure shows the aspects coaches consider when they decide whether or not to make training adaptations and what adaptations to make.

5.4. DISCUSSION

The purpose of this study was to examine coaches' current perceptions about fatigue and their methods of monitoring it during a training session in competitive swimming. As only one other study has investigated a similar concept (Taylor et al., 2012), this study was intended to be exploratory. In the present study, it was found that coaches had a range of familiarity with certain concepts of fatigue in training, including mechanisms, effects and additional factors. In addition, the coaches used a range of methods and equipment to monitor fatigue, with preferences for methods that were familiar and easy to implement, such as stopwatches and visual observation. The variations in knowledge and practices were reflected in a range of beliefs and perceptions amongst the coaches about fatigue, the training process and their role as a coach amongst the responders. Part of this may be due to the complexity of the topics of fatigue and training.

The questionnaire was subdivided into three sections that will be discussed here. These sections are as follows: coaches' knowledge of the topic of fatigue; the methods coaches utilise to monitoring fatigue; and the management of fatigue during training by coaches. Due to the large amount of data and information produced by the questionnaire, only the major findings identified by the coaches, and relating to the overall research topic, are discussed here.

5.4.1. Coaches' Knowledge of the Topic of Fatigue

This section covered areas concerning coaches' knowledge of fatigue, including the mechanisms that cause fatigue, the effects fatigue can have, and finally any additional factors that may influence the fatigue experienced by athletes.

There appeared to be a wide variation in the coaches' depth of knowledge of fatigue during training. Coaches appeared to have high familiarity with some mechanisms, effects and additional factors of fatigue and less familiarity with others. Coaches appeared more familiar with physiological mechanisms and effects in comparison to psychological or biomechanical ones. The variations in coaches' familiarity with certain factors could be an effect of the research on the topic of fatigue itself. This topic is renowned for its difficulties in terms of numerous mechanisms, effects, and methods of measurement, as well as the need for specificity and individual differences amongst athletes as described in Chapter 2 (Enoka and Duchateau, 2008, Phillips, 2015). To attempt to minimise this in the present study, a definition of fatigue was provided at the beginning of the questionnaire (defined in Chapter 2 as *'the inability to sustain maximal swimming velocity'*, Alberty et al., 2009; p. 638). This suggests that the variations amongst coaches could be due to a limited understanding of the topic of fatigue during training. Similar variations in coaches' knowledge and a lack of consistency has also been shown in other studies analysing the content knowledge of coaches on technique identification and that discrepancies are present between coaches in terms of the terminology and rationale of the technique used (Grant et al., 2012). As individuals, coaches come from a wide variety of backgrounds with individual and potentially wide ranging beliefs, experiences and knowledge (Nash, 2008). The knowledge individuals have, and develop, can be a consequence of these perceptions and beliefs. In turn, knowledge and beliefs can influence the actions coaches can use to implement training protocols and monitoring methods (Kirk et al., 2006). This is the first study to analyse the content knowledge of coaches on the topic of fatigue.

Although only some aspects of fatigue were found to be statistically significantly associated with the experience of the coaches, a trend did appear to exist between the highest swimming qualification a coach possessed and the level of familiarity with aspects of fatigue. Those with a higher qualification level stated they were at least 'familiar' to' extremely familiar' with all the mechanisms and effects of fatigue, whereas those coaches with a lower qualification level showed a higher variation in the aspects of fatigue they stated they were and were not familiar with. One reason for this trend may be the sources that coaches stated they obtained their knowledge from. Ninety percent of the coaches stated they had mainly gained their knowledge from experience, which is in line with previous literature that explains experience as an important and realistic aspect of learning vital coaching skills which are essential to improve an athletes' performance effectively (Dorgo, 2009, Gould et al., 2002, Nash and Sproule, 2009). The second main source of knowledge was perceived to be from coach education courses (72%). Although this value was high, it did contradict some of the previous literature that stated coach education courses are limited in their ability to develop knowledge. Formal courses are thought to not provide the knowledge required for coaching, especially for elite coaches where it has been noted that formal courses do not provide sufficient information to develop their knowledge (Dorgo, 2009, Jones and Turner, 2006, Nash and Sproule, 2011, Nash et al., 2011). The coaches in the current study added to this by stating they believed there was a lack of information on the topic of monitoring fatigue during training:

"Don't have enough information on this, more of a hunch than anything else." – Coach 91

There was also a perceived lack of courses or on-going training on this topic:

"Lack of CPD training on this subject via the Amateur Swimming Association (ASA) and Institute of Swimming (IOS)." – Coach 75

This appeared to be related to the perception amongst the coaches that knowledge of this topic was dependent on the level of swimmers coaches were working with. This could be due to the perception that lower level swimmers do not experience fatigue or that they do not need to be monitored:

"*I found this pretty hard to answer, maybe due to the level of swimmer that I work with and my own knowledge levels.*" – Coach 23

This may explain why there appeared to be a trend for coaches with higher qualification levels to source their knowledge from a larger variety of sources in comparison to their level one counterparts, and also to be familiar with more aspects, as those coaches with a higher qualification level usually worked with higher level athletes. Some coaches however, did not deem this necessary and found their knowledge sufficient, regardless of their qualification level, stating they thought all the topics were 'self-explanatory' and 'not-necessary'. Given the range of information on fatigue in sport (Ament and Verkerke, 2009), the many unanswered questions that still exist (Phillips, 2015), and the progression of new findings in research every year this seems a very limiting approach. This could have serious implications for athlete development as important details may be missed or swimmers may receive incorrect feedback which could ultimately be detrimental to athlete development.

Finally, the coaches' perceptions of fatigue may also have an influence on their interest and progress in the development of their knowledge of fatigue. Research has suggested a relationship between beliefs and actions (Kirk et al., 2006). The coaches who took part in the present study showed differing beliefs regarding fatigue and its monitoring during training. Some insisted it is just a part of the sport that the athletes must deal with:

"*You get tired in a race, you have to learn to deal with it.*" – Coach 33

This may be from the observation of the effects of fatigue that have occurred during races and competitions (Smith et al., 2002). Research has emphasised this and highlighted the occurrence of fatigue during races with changes in physiological, biomechanical, and biochemical parameters (Alberty et al., 2009, Ament and Verkerke, 2009). However, it is important to note that swimmers complete higher volumes at high-intensity efforts during training on a daily basis (Arroyo-Toledo et al., 2013, Richmonda et al., 2015). Others insisted that fatigue monitoring was a part of the athletes' development and necessary for them to progress;

"Unless the swimmer experiences fatigue s/he can't learn how to deal with it, take steps to avoid it, recover from it." – Coach 10

This premise regarding the necessity of fatigue is also true as to induce supercompensation and improvements in performance, swimmers need to be stressed to disrupt homeostasis and cause adaptations (Meeusen et al., 2013). Similarly, another coach stated:

"The athletes' response to fatigue and education of fatigue is an integral part of their development." – Coach 83

Finally, some coaches deemed it important for their own development and understanding of the training process as a coach.

"Better understanding of the effect of fatigue and how to create the correct level of fitness to cope with the demands of the training session is very important." – Coach 62

One factor which may have influenced coaches' perceptions of fatigue may have been the questionnaire itself and simply asking the coaches to complete the questionnaire. Enquiring about coaches' understanding, the methods they use and the importance placed on fatigue and its components may have changed how coaches answered each question. Although coaches were prevented from returning and changing their answers to previous questions and were asked to answer the questions honestly and individually, it could not control how coaches answered each question nor any changes in coaches' perceptions of fatigue throughout the questionnaire.

The role of education and the information coaches may initially receive on this subject may also explain the variation found amongst this group of coaches. Part of this information comes from sport science knowledge and research which is deemed to form a significant part of the knowledge base required by coaches to effectively complete their role (Martindale and Nash, 2013). However, it has been noted that there are difficulties transferring knowledge from sport science to the coaching environment (Martindale and Nash, 2013). This, in addition with the difficulties of the topic of monitoring fatigue, may mean that coaches are not obtaining adequate information in terms of fatigue and its management during training and is emphasised by the differing results obtained from the present study. This is important as if coaches do not know this then relevant information cannot be fed back to athletes to continue performance development. As a result, the differing beliefs and knowledge about fatigue may also influence the actions of a coach during the training process. The variations amongst coaches in their beliefs and knowledge also show the potential lack of consistency in the education of this topic and lack of clarity which presently exists in coaching. This indicates a need for a more structured and effective education of this topic amongst coaches, continued CPD and updates or disseminated information from studies such as this as new results emerge, and applied information that can be utilised by individuals of all coaching levels.

5.4.2. The Methods Coaches Utilise to Monitor Fatigue

The questions asked in this section referred to the methods and technology or equipment coaches could use to monitor fatigue during a training session. One of the first results from this section was that although a high percentage of coaches stated they monitored fatigue, not all coaches did. Due to the vital role fatigue plays in athlete progression, training design, skill development and automation, this was surprising (Bonacci et al., 2009, Meeusen et al., 2013). Of those methods identified, coaches predominantly stated they used those methods which were quick to provide information, easy to utilise, and reliable. These included visual observation, stopwatches, and above-water cameras. These methods were also used most frequently, ranging from every session to every week. The main reason coaches stated for not utilising other methods available to monitor fatigue was accessibility, including: time; access to the equipment; the cost of the equipment; understanding of the equipment; and the quality of the results it produces. Similar findings were identified by Taylor et al. (2012) who noted that the rationale for the lack of use of certain equipment by individuals working with elite athletes in New Zealand was the expense of certain equipment or that they required too much time or support. Although this study investigated those individuals working with elite athletes, swimmers at a range of performance levels follow the traditional training periodization used in swimming consisting of multiple high-intensity training sessions per day (Richmonda et al., 2015). Further research is needed to determine if similar perceptions, methods and issues are present in coaches who work with swimmers at other performance levels.

A trend was identified in the present study among coaches with higher qualification levels and experience appearing to use a wider range of equipment to monitor fatigue, and often also using this more often than their colleagues. This was emphasised in the types of equipment statistically significantly associated with years coaching or qualification levels. Certain methods were also discovered to be used by all coaches, regardless of the coaching experience. These included visual observation, stopwatches and noting the athletes' personality or mood. Taylor et al. (2012) also noted that those monitoring fatigue in athletes in New Zealand used a self-report questionnaire most often. Thus the coach-athlete relationship appears to be a major factor in monitoring fatigue and training intensity. Coaches in the present study agreed with this, commenting that:

"As you get to know your athletes I think that by far the best analysis tool is simply talking to your athletes." – Coach 36

In addition to the coaches' experience, the use of equipment to monitor fatigue appeared to vary again in terms of coaches' perceptions of what was adequate for the level of their swimmer. One coach stated:

"I feel that for the age and level of swimmer I am currently coaching, the tools I use are appropriate." – Coach 35

Whereas another coach stated they would use any means or methods relevant to enable an improvement in their athletes:

"Yes I will use whatever I can to increase my knowledge of the swimmers." – Coach 64

The variations in responses and lack of consistency again note a lack of clarity amongst coaches in the most effective tool and method to use to monitor fatigue during training in swimmers.

Due to the limitations arising and lack of use of many methods of monitoring fatigue, it was of interest to understand what methods coaches feel are required to develop the monitoring of fatigue during training. Therefore, coaches were asked what type of equipment they would like to obtain to monitor fatigue in the present study and stated: heart rate, underwater cameras and blood lactate analysis. Coaches stated they used these methods for the purpose of monitoring the training load or intensity, the technique during training and to more effectively feedback information to athletes. Although less of an issue at elite level, one major premise in the present study was applying this to a group of athletes rather than one or two individuals with 'too many swimmers at a time':

"Try to individualize as much as possible, but if you have 20 athletes in three lanes it becomes difficult." – Coach 80

To use these methods and pieces of equipment to monitor fatigue in a group of swimmers would involve overcoming some of the issues the coaches identified as being related to using equipment to monitor fatigue, including: difficulties understanding the equipment, ease of accessibility, cost and time consumption for use.

Again, despite a variety of available methods to monitor fatigue and accessibility by coaches now improving as a result of technology development, coaches are still resorting to traditional methods of observation and knowing their athletes. The main reason provided for this was the limitations of accessibility of many pieces of equipment into a training environment. Within the methods coaches use to monitor fatigue there is also a variation amongst coaches and only one method on which all the participants of this survey agree as a tool for monitoring fatigue; visual observation. Part of this appears related to their knowledge and understanding of fatigue, partly a result of their coaching experience but mostly the belief that coaches are limited by the accessibility and application of standard methods of monitoring fatigue.

5.4.3. The Management of Fatigue During Training by Coaches

This section concentrated on what coaches do to prevent or manage the fatigue their swimmer experiences during a training session. A very high majority of coaches made adaptations to their session plan in order to enable their athletes to cope with the intensity of training, with a focus on the individual and rest. One coach insisted that adaptations were all about the individual:

"*All swimmers are individual therefore I must look at all factors of an athlete to make the right choices.*" – Coach 29

However, some coaches noted an issue with this stating that some athletes can take advantage of this perspective and attempt to fool the coach:

"*Only during extreme circumstances as athletes are able to change their behaviour to appear to be fatigued when maybe their mental focus is not where it should be and in doing so taking control of the session by underperforming and 'fooling' the coach into making the session easier.*" – Coach 6

This related very strongly to the coach's relationship with the athlete and that, as in the methods utilised to monitor fatigue, knowing the athlete was also vital in the process of the management of fatigue:

"*Knowing the swimmers allows me to change and make adjustments during training knowing the individuals strengths and weakness.*" – Coach 2

This relationship included the need to guide the swimmers, aid their development and understand them during the process of training and performance development. One coach neatly summarised this by stating the two-way process that the identification and management of fatigue involves:

"*They need to learn for themselves what fatigue is, identify it, the symptoms and the dangers, effects and possible injury implications it can cause. So it becomes a two-way process. Not just me as the coach watching for the warning signals. But they can approach me and let me know*" – Coach 95'

This suggests that this coach is attempting to create and promote a very motivating and educational environment for their swimmers and are accepting of the fact that they, as a coach, and fatigue have an important role to play in this process. However, not all coaches' perceptions were the same and this clearly influenced their actions when it came to the management of fatigue. One coach stated:

"*My motto is "What doesn't kill you makes you stronger.*" – Coach 86

This appears to indicate a perspective that athletes need to be pushed through fatigue and it is a necessary part of training. Thus athletes will be pushed to train farther and harder regardless of the outcome. These contradicting perspectives of fatigue could result in very different training environments and consequential effects of training for the swimmers. Another coach had the opinion that a certain level of fatigue was simply detrimental to performance:

"*I strongly believe that if a swimmer is fatigued beyond the point of not being able to maintain their target times then they are too tired to train effectively.*" – Coach 22

This appears to indicate that the coach is more interested in the swimmer being able to train effectively to induce results. These differing opinions are a common rift that currently exists in the design and implementation of training sessions between coaches in swimming. It is the on-going debate regarding quality training over quantity training and the presence of these two opinions amongst a number of coaches in the present study insinuates that this debate is still on-going. Due to the varying content knowledge and methods of monitoring fatigue identified in the previous two sections, it does not seem surprising that variations between coaches continue into their management methods of fatigue during training and the outcomes they attempt to achieve through training. Further research into the training methods, fatigue monitoring methods and the application of these in the management of training is required to help understand these processes. Further research is also needed to assess the methods coaches currently consider effective, such as visual observation.

5.5. CONCLUSION

This study successfully explored coaches' current perceptions about fatigue and their methods of monitoring it during a training session in competitive swimming. It identified that coaches' knowledge, methods of monitoring and managing fatigue differed amongst coaches, particularly in relation to the coaches' experience. This was a topic in which coaches were interested but noted a lack of information regarding its application in terms of coach education programs and information. Thus coaches tended to utilise methods that were quick and easy to use and implement to a squad of swimmers, specifically visual observation and knowing their swimmers. One limitation of the present study, due to its exploratory nature, was that it did not assess the knowledge of coaches but simply determined areas with which coaches were familiar. Now that certain areas have been identified, a necessary future study would be to test the level of content knowledge and assess what coaches actually know about fatigue; its mechanisms, causes and influences. In addition, the veracity of the results coaches are providing must be established as the questionnaire topic itself and completing the questionnaire may have altered coaches' answers and perceptions of fatigue. Finally, it is important to assess whether the method coaches stated they were currently using to monitor fatigue during training, namely visual observation, is effective. The varying opinions, perceptions and practices of coaches throughout this questionnaire indicate a lack of consistency of this topic and thus further research is essential to ensure coaching methods and protocols are effective, and continue developing coaching knowledge and athlete performance.

Chapter 5: Summary

What was already known about this topic?

- Fatigue is a necessary component of the training process.
- There is a lack of literature on the topic of the management of fatigue in training, specifically in relation to coaching.
- Elite coaches utilise self-report questionnaires and practical maximal tests to monitor fatigue during training.
- The application of sport science methods of identifying fatigue are difficult in a training environment.

What new information does this chapter provide?

- A large number of coaches stated they monitored and adapted training sessions based on the individual's ability to cope with the intensity of training and rest durations.
- The coaches perceived they were familiar with a number of aspects of fatigue in training, including mechanisms, effects and additional factors.

- There was a lack of consistency amongst coaches in terms of their perceptions of fatigue during training and the methods they used to manage this.
- Coaches stated they used a range of methods to monitor fatigue during training.
- The knowledge and methods used by coaches were mainly obtained through experience.
- Coaches stated they used visual observation, stopwatches, and selfquestionnaires most frequently, regardless of coaching experience, to monitor fatigue.
- The lack of use of other methods of monitoring fatigue was perceived to be mainly a result of a lack of accessibility.

6

An Evaluation of Swimming Coaches' Observations of Technique Adaptations during Training

6.1. INTRODUCTION

The role of the coach is widely accepted as multifaceted and can differ according to the performance level of the athlete, the context of the coaching programme and the sport itself (Szabo, 2012). This has resulted in many attempts to define the role of the coach in a number of coaching contexts and sports (Horton, 2015, Szabo, 2012). According to Horton (2015), there is no consensus as to the definition of a coaches role as every coach and coaching context is different and as a result is very difficult to define. As all of these elements of practice are bought together during a training session, designed and implemented by the coach (Nash and Sproule, 2011, Smith, 2003), the training session is considered a critical element in the development of performance (Hodges and Franks, 2002). It is therefore extremely important that the information collected during training and athletic performance is objective, unbiased, accurate and as comprehensive as possible (Hughes and Franks, 2007).

A key factor in the training process is the feedback the athlete receives, and is defined as *'performance related information that the learner receives during and after performing the task'* (Pérez et al., 2009; p. 30). This information can be both intrinsic and extrinsic. Intrinsic feedback is that which is available to the learner directly, by virtue of the skill or environment in which the skills is performed (Magill, 2004). Extrinsic, or augmented, feedback is from an external source (Magill,

2004). Figure 6.1 has been adapted from one suggested by Bishop (2008) to portray the coaching feedback cycle.

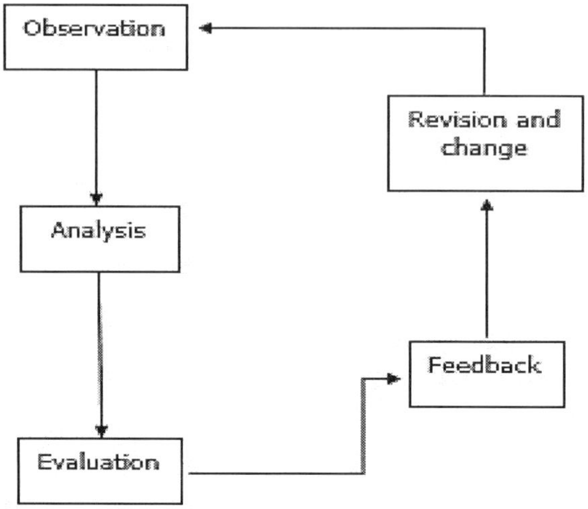

Figure 6.1: The coaching feedback cycle: an interpretation (Bishop, 2008b).

Hodges et al. (2003) stated that athletes learn and perform skills better when they receive extrinsic (augmented) feedback than only intrinsic feedback. Extrinsic feedback must be purposefully applied and thus a number of factors can influence its effectiveness, including: the level of feedback, the amount of feedback, the timing of feedback, and the skill of the individual receiving the feedback (Hodges and Franks, 2002, Magill, 2004). Large amounts of feedback may be beneficial early in the learning process, but too much feedback later in learning may actually impair performance. High frequency feedback may result in a dependence on that feedback so that in competition situations the athlete's own error detection ability and self-correction mechanisms are not activated (Phillips et al., 2013). There is also evidence that over-guidance can restrict the creativity of athletes during performance. Finally, if the level of feedback is too specific for the level the performer, the feedback could hamper performance and hinder outcome success (Phillips et al., 2013). The feedback athletes obtain extrinsically is usually determined by the coach, and in competitive swimming is usually provided verbally on poolside (Pérez et al., 2009).

In order to provide feedback which can be incorporated into planned practice (training), a coach must first observe and analyse the performance (Hodges and Franks, 2002). This involves a number of skills, including: a thorough knowledge and understanding of how to effectively analyse strokes, diagnose strengths and weaknesses, articulate desired movements and increase swimmer's efficiency, as well as communicate information on optimal movement patterns or errors (assuming they are observed correctly) back to the swimmer in an effective manner (Hodges and Franks, 2002, Leas and Chi, 1993, Waters et al., 2014). This involves the need for coaches to plan and organise learning experiences, provide instructional information, and thus coaches must rely on a very high level of perceptual-cognitive skill, which enables the coaches to integrate visual information with existing knowledge (Hodges and Franks, 2002, Mets et al., 2003, Waters et al., 2014).

The importance of observation for coaches and its application in the evaluation of athletes' performances has been noted in the literature by many scholars (Bird and Hudson, 1990). This research has predominantly been completed by two authors, Barrett (1979), Barrett (1983) and Allison (1987). Barret focused on the need to plan what coaches observe and how they focus their attention on the identification of critical features. Alison investigated whether this differed between elite and novice coaches (Bird and Hudson, 1990). Part of the ability to judge the quality of human movement and provide the most appropriate intervention to improve performance relies on the capacity to observe, recognise and estimate critical features accurately, as well as discard factors which are judged to be irrelevant (Hernández et al., 2006, Morrison et al., 2005, Mets et al., 2003). The relevance of feedback that a coach passes onto a swimmer is therefore directly related to what they have observed (Mets et al., 2003).

The coaching process is not flawed, but the observation and analysis phases of the process are limited. Traditional coaching often involves subjective observations and conclusions based on the coach's perceptions, biases and own previous experiences. Many coaches are able to anticipate events and make appropriate changes to influence performance, but even the best are prone to human error and making wrong decisions. A number of studies have revealed that subjective observations are potentially both unreliable and inaccurate (Franks and Miller, 1991). Coaches are often only able to recall up to 50% of key performance factors, even with special training in observation, and reproducibility of observations can vary amongst coaches (Franks, 1993, Franks and Miller, 1991, Mets et al., 2003). It was also pointed out as an additional finding by Franks (1993) that coaches not only made wrong assumptions when events did not exist, but they also stood by their false pre-conceived ideas. Several authors have also examined the quality of coaches' visual observations of sport actions and identified that coaches' estimations of kinematic angles and velocities during running scenarios, and range of motion of joints in vertical jump evaluations was poor (Knudson, 1999, Krosshaug et al., 2007, Morrison et al., 2005). An inability to correctly recall or identify sporting actions may result in the relay of incorrect or insufficient feedback to the athlete, potentially preventing development. These studies show that the observations processed by normal human memory are very unreliable as a source of information for feedback of performance (Franks, 1993). During a training session, these flaws in accuracy of observation are exacerbated by the necessity to observe a large number of individuals for long periods of time. Therefore, improving the accuracy of observation is very important to feedback and athlete development.

Accuracy of observation of technique during swimming is affected by certain technical actions being obscured by water, a medium which is 1000 times denser than air (Leas and Chi, 1993, Maglischo, 2003) and therefore can distort the image through refraction and turbulence. This may be even more difficult during training due to the large number of swimmers making the water turbulent and impairing visibility. As a result, the ability of competitive swimming coaches to observe technique during training may be limited.

A small number of studies have investigated coaches' visual observations of swimming in terms of technique performance or the detection of stroke errors (Hannula, 2003, Leas and Chi, 1993, Moreno et al., 2006, Persyn and Colman, 2005, Waters et al., 2014). However, these studies had small sample sizes of coaches (n=4), focused only on frontcrawl, predominantly used eye tracking systems and did not assess these factors during a training scenario. This is important as, even with the eye-tracking system, the common incidence of poor visibility of actions under the surface of the water during training remains an issue. In addition, although high-intensity training has been shown to result in technical changes (Chapter 4) and coaches are predominantly using visual observation as a method to monitor these technical changes during training (Chapter 5), no research has analysed this. Due to the impact of high-intensity training on technical performance, the importance of technique for swimming performance and the role of effective feedback in this process, it is very important to understand what aspects of technical performance coaches are currently observing during training to monitor their swimmers.

The role of feedback is central in the performance improvement process, and by inference, so is the need for accuracy and precision in the feedback. To improve the communication of feedback to athletes, coaches have often attempted to use alternative quantitative aids to provide accurate and precise extrinsic feedback. Video is one alternative method which is growing rapidly in popularity for use during training due to its relatively low cost, accessibility and portability (Hughes and Franks, 2007, Liebermann et al., 2002). The evaluation of observation simplified by the use of video was first assessed during the 1970's (Johansson, 1975) and it was found then that the more experienced the coach, the more detailed information they could provide from the video (Liebermann et al., 2002). Imwold and Hoffman (1983) continued this research by conducting a series of studies about the evaluation of observation with a specific biomechanics focus, similar to the scope of the present study. Imwold and Hoffman (1983) and Liebermann et al. (2002) both found coaches were able to provide plausible technique descriptions using video and the ability to link those descriptions to observations was found to be dependent upon the experience of the coach and their familiarity with the skill.

Analysis based on accurate observation and recall during training is a key tool for improving future performance for two reasons (Bishop, 2008a). Firstly, when a coach does not perceive detail, their feedback will be reduced in quality and may not include information useful for the learning process (Imwold and Hoffman, 1983). Therefore, it is vital that the information coaches observe is accurate and adequate. Secondly, athletes depend on this information for performance development. The provision of effective feedback can only be facilitated if performance and practice involves a vigorous process of analysis (Hughes and Franks, 2007). Further research is needed to assess whether the methods coaches are currently using to monitor technique and fatigue during training are effective in swimming. It is also important to determine whether the video can aid these methods.

Although advances in technology have made it possible to augment and improve feedback to athletes during training, and this is often deemed invaluable, the application of research has not advanced as rapidly (Liebermann et al., 2002). Coaches still rely on qualitative analysis through a combination of visual observation and simple timing data to evaluate technique and athletic performance, respectively (Fleming et al., 2010, Mets et al., 2003). This highly subjective approach is limited to the athletes' and coaches' interpretations of observed actions that can last a fraction of a second and its accuracy has been shown to vary widely depending upon the coach. To date, the current study undertaken is the first study to investigate coaches' visual observations of technical changes during training in competitive swimming and assess whether video feedback can improve this process in coaches.

The coach plays a vital role in athlete feedback and development during training, including the coaches' management of fatigue. However, little is known about the factors coaches observe, nor whether they can observe the technical changes which have been identified as occurring as a result of fatigue. To work effectively, coaches need to know about technique errors to be avoided, as well as stroke technique that should be guided. To aid coaches to develop athletic performance, collaboration is required to link biomechanical knowledge with observational abilities. This can aid coaches that rely on observational skills to develop simple and effective plans to analyse the movements of their performance (Bird and Hudson, 1990). This is important not only to continue athlete development, but to understand how observational skills of coaches can be developed (Leas and Chi, 1993).

With respect to assessing the level of fatigue of a swimmer, the coach needs to be aware of the variables that should be observed and to become skilled in observing them (Hudson, 1990, Martindale and Nash, 2013). To date, there has been a lack of investigation into coaches' visual observations of technical changes during training and whether video feedback can improve this process.

6.1.1. The Purpose of the Study

Therefore, to investigate whether the use of technical indicators of fatigue in breaststroke swimming can aid competitive swimming coaches' abilities to visually observe and identify the changes in technique associated with acute fatigue during a training session, the aims of this study were to:

- Identify the technical indicators coaches currently observe, if any, to indicate fatigue during training.
- Determine whether coaches can observe and identify the changes in technique (identified in chapter 3) which occur as a result of acute fatigue using visual observation.
- Investigate the effect of an intervention and retention test on competitive swimming coaches' abilities to observe and identify the changes in technique which occur as a result of acute fatigue during training conditions in breaststroke swimming.

It was hypothesised that:

- Coaches will identify a range of technical indicators they use to observe their athletes during training.
- The technical indicators coaches will use will coincide with the technique breakdown used in coach education of: Body, Legs, Arms, Breathing, and Timing (BLABT).

- The technical factors coaches observe will be related to their perceptions and beliefs about fatigue during training.
- Coaches will be able to identify those technical variables visible above the water but will be unable to view those taking place during stroke phases below the water.
- A training intervention will improve the coaches' abilities to observe and identify the acute changes in fatigue occurring during training conditions.

6.2. METHODS

6.2.1. Participants

Twenty practising competitive swimming coaches took part in this study. They consisted of six females and fourteen males, aged 39.7 ± 13.1 years. All coaches had several years coaching experience, ranging from 3 to 40 years (average years coaching 11.2 ± 9.5 years); and coaching qualifications ranging from level 1 to level 4. Coaches spent 13.9 ± 8.1 hours coaching per week at 7.1 ± 3.1 sessions a week. To be eligible in the present study, participants were required to have completed the questionnaire in Chapter 5 and be currently coaching national age-group performance level, or higher, competitive swimmers in Scotland. The participants' demographic information is detailed in Table 6.1.

Table 6.1: A description of the participants' characteristics.
F = female, M = male, C = control group, E = experimental group

Coach	Age (years)	Coaching quali-fication level	Years coaching experience (years)	Gender	Total hours coaching water sessions (hours)	Group Allocation
1	46	3	20	F	20	C
2	42	2	3	M	6	C
3	26	3	8	M	10	E
4	21	1	3	M	16	C
5	68	3	40	M	12	C
6	37	3	15	M	36	E
7	47	3	15	M	18	E
8	36	1	3	M	2	E

Coach	Age (years)	Coaching quali-fication level	Years coaching experience (years)	Gender	Total hours coaching water sessions (hours)	Group Allocation
9	53	4	27	M	15	E
10	46	2	13	F	3	C
11	22	2	8	F	10	E
12	33	3	10	M	18	C
13	26	3	5	M	25	C
14	25	3	10	F	18	C
15	59	3	18	M	20	E
16	49	3	8	F	10	E
17	53	2	5	M	10	C
18	26	3	5	M	16	E
19	39	1	4	M	4	C
20	39	2	4	F	8	E

Prior to the test session participants were provided with an information document, detailing the purpose of the study, the procedure of data collection, and the potential risks and benefits of being involved in the study. They provided informed consent, and completed the participant information form (see Appendix 9). The study was approved by the Edinburgh University Ethics Committee.

6.2.2. Experimental Design

This study's design was a pre-post, randomised-groups design, with a retention test. This design was used for three reasons. The first was to determine whether coaches could observe and identify a larger number of technical changes after receiving a video-based intervention. The second was to determine whether any observation skills developed during the video-based intervention were retained or forgotten. The third was to ensure that the two groups were randomly formed in an effort to control for sources of invalidity (Thomas et al., 2011). The study took place during the early preparation phase of the swimming training season as it was believed coaches would have more time available to participate in the study due to fewer competitions during this time (Maglischo, 2003).

Data collection consisted of five stages which involved three test sessions:

- Stage 1: Participants were randomly allocated into two groups, a control group (C) and an experimental group (E) (n=10 in each group), using a block randomisation technique. This was to ensure that each group was formed randomly and with an equal number of participants (Field, 2009).
- Stage 2: Test session 1, using video footage of Swimmer 1.
- Stage 3: Depending upon their group allocation, each participant received a video-based training intervention (group E) or a break with no treatment (group C) (stage 3) lasting 1 hour.
- Stage 4: Test session 2, using video footage of Swimmer 2.
- Stage 5: Following a 4 week break, test session 3, using video footage of Swimmer 3.

These five stages are represented schematically in Figure 6.2.

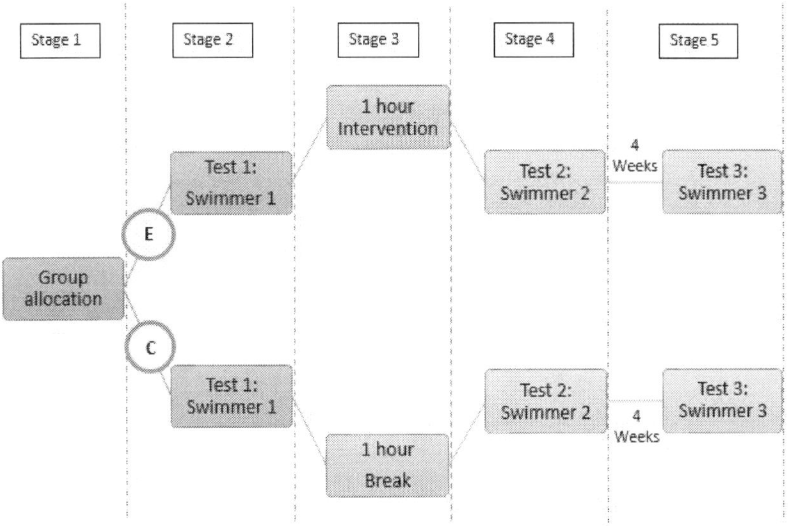

Figure 6.2: The research design and stages of data collection. This schematic displays the five stages of data collection, three test sessions and group allocation. E = the experimental group, C = the control group, Swimmer refers to each swimmer viewed during each test session (n=3).

The first two test sessions took place on the same day, separated by a period of one hour. Although, according to Vargas-Tonsing (2007) and Erickson et al. (2008), coaches are interested in coaching education, particularly when the topic is of interest, there was a concern that coaches would be unavailable to complete the intervention due to other coaching commitments. It was therefore decided that that intervention would last one hour to maximise the coaches' participation and due to the exploratory nature of this research. Previous studies looking at sport and coaching have also used interventions ranging from one to two hours (Smith et al., 2007).

After completing the second test session (Stage 4), each participant, in both groups, had a four week break from the study, during which time no technical training was provided, before performing the final retest session (Stage 5). Magill (2004) indicated that four weeks was a sufficient time to determine whether the skills learned during the training intervention were retained or forgotten by the experimental group.

Each test session (performed in Stage 2, 4 and 5) lasted a total of thirty minutes and involved the same data collection protocol. The participants were made aware that their technique observations skills were being analysed but were blinded to the use of different groups and treatments being undertaking. The participants were fully disclosed this information at the end of the study. Each participant attended their test session individually to prevent the data collected being influenced by the presence of other participants and were asked to refrain from talking about their sessions until the end of the data collection phase to ensure those participating later in the study remained blinded to the differing groups and treatments. Ensuring participants were blinded to the use of different groups and treatments was important to protect the internal validity of the study and minimise expectancy and avis effects from the participants (Thomas et al., 2011). This was used in an effort to analyse the effects of the intervention and attempt to prevent history and maturation effects on the validity of the data (Thomas et al., 2011).

6.2.3. Data Collection Methods

The coaches' task was to observe and evaluate the technical performances of three competitive swimmers from above-water video clips. Each swimmer had two video clips: the first video clip displayed the swimmer completing a 100m breaststroke swim in a fresh state. This was to obtain a neutral image of the swimmer's natural technique; the second video displayed the same swimmer completing a 100m breaststroke in a fatigued state, after the swimmer had completed a high-intensity training session. Each video clip showed the 100m swim from three above-water perspectives, one side view and two front view perspectives. This was to mimic coaching under training conditions, in which coaches observe from an above-water perspective (as identified in Chapter 5). Only one swimmer's video clips were observed in each test session and all coaches viewed the same video clips at each test session. The swimmer to be observed in each test session was selected using a simple randomisation technique of flipping a coin (Thomas et al., 2011). The swimming performances were videoed prior to the coaches' test sessions using three above-water cameras (Panasonic VC-100, Panasonic Corporation, Osaka, Japan), two in front of the swimmer's path (at either end of the swimming pool) and a third from a side perspective, at the centre of the pool, parallel to the swimmer's path. Each above water camera was positioned at a height of 1.5m above the water in an attempt to mimic the coaches' normal visual perspective on poolside. This camera setup was similar to that used in Chapter 4. The swimmers filmed for the video clips were elite, international level swimmers and had participated in the study completed in Chapter 4. The quantification of technique changes in these swimmers were established using the same methods employed in Chapter 4.

Each test session took place in a location of each coach's choosing which was isolated, quiet, and free from distraction (Nash and Sproule, 2012). Before beginning each test session, the two pre-recorded video clips of the relevant swimmer were prepared for viewing by the experimenter using Dartfish video analysis software (Dartfish Ltd, Fribourg, Switzerland). This ensured the coaches simply had to press play or pause to observe each video clip. Each coach was given an instruction and data sheet (see Appendix 10), briefed about the protocol, and provided with the pre-prepared laptop (Dell Inspirion, Dell, Texas, USA) at the beginning of each test. Coaches were seated a half metre away from the laptop which had a 15inch screen. A laptop screen was used as this would be the size of screen (or smaller) coaches would utilise if they were using an above-water method of video collection on poolside. The laptop screen displayed all three camera perspectives using the split image tool on the software and were synchronised to play at the same time, as shown in Figure 6.3.

Once ready, each coach was allowed to observe the two videos of the swimmer from all perspectives at the same time. Participants were only allowed to view each video once, and could not use the slow motion or rewind functions but were allowed to pause the video to make notes or specify any observations. Again, this was to mimic coaching conditions in which coaches only watch a swimming trial once (as identified in Chapter 5; Maglischo, 2003).

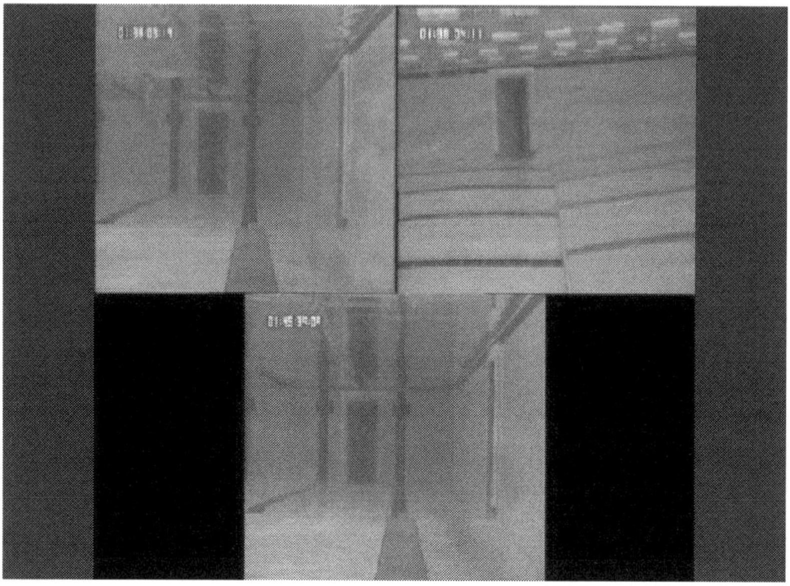

Figure 6.3: The combined camera perspectives observed by the coaches.

As one of the aims of this study was to explore coaches' observations of technique when the swimmer was fatigued, and the use of specific keywords may lead the coaches in their answers or observation focus, participants were not provided with any information regarding what they should be observing (Wulf and Prinz, 2001). Instead, coaches were asked to observe the swimmer's technique and diagnose any changes in the swimmer's technique between each video, making notes of these observations using the data sheet provided (see Appendix 10). This included asking the coaches to be as specific as they could, and to include both positive and negative observations (Leas and Chi, 1993).

On completing their observations, each coach was briefly interviewed and asked a series of sixteen questions (Appendix 11). The order of questions varied between the control group and the experimental group in order to keep participants blinded to the use of two groups and an intervention. However, by the end of the final test session, all coaches had been asked the same questions and were fully briefed regarding the full research purpose (Thomas et al., 2011).

The interview was designed to explore the coaches' observation skills and use of video, for both feedback and analysis during training as well as probe coaches' knowledge of fatigue during training. Consequently, it was semi-structured, with the purpose being to investigate the coaches' views on the technical markers and their use in visual observation of fatigue during training, focussing on their own experiences and expertise. This form of interview was considered the most appropriate as it made allowances for unanticipated issues or questions which may arise (Patton, 1990). It also allowed for continual probing and checking of responses from each coach on the topic of fatigue and its impact upon technique during training. The coaches could respond for as long as needed for each question and all the responses were audio recorded. The questions for the interviews were constructed by the lead researcher and arose deductively from the data gathered from the studies in Chapters 4 and 5 (Nash and Sproule, 2009). The questions were then discussed with a second researcher to ascertain their appropriateness and potential to elicit responses to the topic under investigation. Please refer to Appendix 11 for the interview schedule.

6.2.4. Intervention Protocol

The training intervention was a single, one hour, video- and computer-based training session, in a professional development format and was provided individually to those in the experimental group. The aim of the intervention was to educate coaches on the topic of monitoring fatigue during training, develop coaches' use of video during training, and improve the coaches' abilities to identify acute technical markers of fatigue in breaststroke swimming. The training intervention was broken down into three phases. First, a discussion took place between the coach and experimenter. During the discussion coaches were allowed to ask any questions and discuss subjects relevant to monitoring fatigue during training and the topics listed below in Figure 6.4. Second, coaches were provided with the results of research into this topic and a brief summary of the previous three chapters of this thesis.

Finally, following this discussion, coaches then underwent a practice session using the Dartfish software. While practicing identifying and measuring each technical variable, coaches were provided with under-water, as well as above-water, camera perspectives. Both perspectives were used in an attempt to allow coaches to relate visually what was happening technically below the water with technical actions above the water. This approach was used in an effort to allow the coaches to apply this knowledge to a coaching perspective.

The intervention was structured yet flexible, similar to the interview design. This was also to allow for any unexpected issues, queries or comments which may have arisen. The training was standardised amongst all participants by ensuring that the amount of practice and information each participant received was constant. This prevented the participants being overloaded with too much information during this time. The same coach provided all of the training throughout the study. To remove experimenter bias and subjectivity the participants were unfamiliar to the coach. The total duration of the intervention, or non-treatment (control) group, was equivalent amongst all participants, being one hour, regardless of their group allocation. During this hour the control group received no training and were to simply continue with their daily routine until the next session.

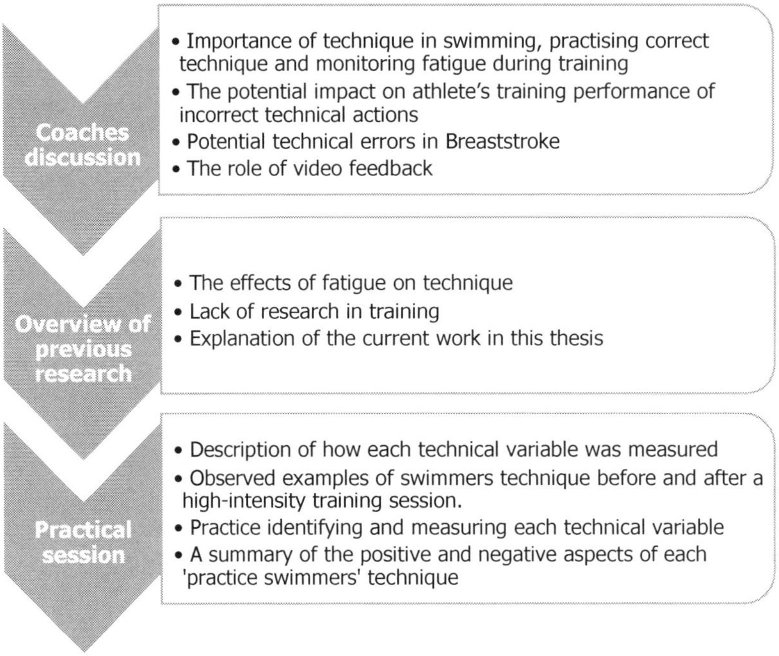

Figure 6.4: The content of the training intervention.

6.2.5. Data and Statistical Analysis Methods

All statistical analysis was completed using SPSS (version 19.0, IBM UK Limited, Portsmouth, UK). The dependent variable (the number of technical changes correctly identified by each coach) was obtained for each coach, from all three test sessions. The independent variable was the form of treatment that the coaches underwent. The data were analysed through two processes; first the coaches' actual observations of technical changes with fatigue was evaluated and the effect of the intervention assessed; secondly the coaches' comments and perceptions regarding the observation and identification of fatigue during training were investigated. These two processes will now be discussed.

6.2.5.1. The analysis of coaches' observations of technique and the effect of a video intervention

The proportion of technical variables correctly identified by coaches was totalled for each participant and also presented as a percentage. The change in the proportion of variables identified between Session 1 and 2, and Session 2 and 3, was then calculated for each individual and presented as a percentage. To assess whether the effect of the video intervention or practice could improve coaches' abilities to observe and identify technical changes related to high-intensity training, the differences within and between groups was assessed at equivalent test sessions. Data were initially assessed for normality using a Shapiro-Wilks test. These tests found the data to be statistically significantly not normal ($p < 0.05$) for the control group ($p = 0.022$, 0.036, and 0.01 for Session 1, 2 and 3, respectively) and for the experimental group for test Session 3 ($p = 0.002$). The remaining two test Sessions were normal ($p = 0.107$ and 0.478 for Session 1 and 2, respectively). The data remained not normal even after attempts to transform the data. As a result of this, non-parametric tests were utilised. To compare differences between each pair of groups for each session, Mann-Whitney U tests were conducted. To test for statistically significant differences within groups and between sessions, Wilcoxon matched pairs tests were used. Any changes in the proportion of technical variables correctly identified were compared from Session 2 and 3 relative to Session 1. To assess the effect of retention, changes in Session 3 were compared relative to Session 2. These two methods enabled the statistical assessment of the effect of an intervention and the retention ability in the experimental group, and of practice and its effects in the control group (Field, 2009). Each test was conducted using the relative change in percentage as sources of input. In addition to the previously described tests, a 95% confidence interval (CI) of the true mean was quantified for each Session to also assess whether any changes in the proportion of variables identified were statistically significant (Hopkins, 2004). The upper and lower CI boundaries were used to indicate the range in which the true value of the change in the variable falls 95% of the time. The same methods were used as those described in Chapter 4 (Portney and Watkins, 2000). This minimized the risk of making both Type I and Type II errors and ensured that the power of the study was maintained (Portney and Watkins, 2000).

In addition, effect sizes were calculated between each session to measure the meaningfulness of any changes for each group. The effect sizes were calculated for each group using the Z-score produced by the non-parametric tests described above and the total number of observations (n) in the following equation: r = Z/Ön (Field, 2009). This method was used due to the non-parametric nature of the tests (Field, 2009). These values were calculated for each group, per session and used to determine group effect sizes for Session 2 and 3 relative to Session 1. This enabled a measure of cumulative learning across the feedback sessions to be obtained. Effect sizes were also calculated for the retest Session 3 relative to Session 2 to assess the effect of practice in the control group, and as a measure of retention for the experimental group. The effect sizes were interpreted according to Hopkins (2002) where 0 is trivial; 0.2 is small; 0.6 is moderate; 1.2 is large; 4.0 is very large; and infinite is perfect, as described in Chapter 4.

6.2.5.2. Coaches' perceptions of the observation and identification of fatigue during training

After each test session, each coach was interviewed and their responses audio-recorded. Each interview recording was transcribed and written verbatim, by the main researcher. The purpose of the data analysis of the transcriptions was to explore the visual technical indicators coaches currently observe to indicate the fatigued state of their athlete and to provide additional insight and richness to the quantitative data collected. The transcriptions were then analysed through the quotes and themes found in the words of the participating coaches using the guidelines of thematic analysis. That is, each transcript was read and re-read repeatedly and any significant statements relating to and illustrating the various dimensions of the core theme of the perceptions of coaches on the observation and identification of fatigue in swimming were acknowledged, selected and noted separately (Nash, 2008). This allowed for the depth and richness of the coaches' responses to be reflected in the results (Nash and Sproule, 2009). Following this, the typed and printed chunks of data were organised into categories and themes determined by the data content (Dorgo, 2009). These were then refined and re-analysed, resulting in the emergence of conceptual categories or themes. Labels were then assigned to the categories/ patterns/themes and when no new categories emerged, it was assumed all had been identified and theoretical saturation had been reached (Nash, 2008). As the transcripts were analysed and organised into themes based on current and existing literature and theories relating to the core theme of the perceptions of coaches on the observation and identification of fatigue in swimming, this process was conducted deductively (or in other words the reality of the coaches' comments was compared to existing literature).

Once clearly distinguishable themes had emerged, the researcher reviewed and (if needed) reorganised data pieces ensuring that all pieces were placed under the appropriate theme (Dorgo, 2009). The categories and patterns that were shown to come together in a meaningful yet distinct way across the transcripts, were developed into the major themes (Nash, 2008). The lead researcher's interpretations of the themes were then reviewed by a second researcher, thus ensuring investigator triangulation (Nash and Sproule, 2012). Investigator triangulation involved the second researcher receiving a sample of typed chunks of transcript data and being asked to independently and deductively attach these data pieces into the main themes. Following this process both researchers discussed the process the main researcher had utilised, using an inquiry audit, to assess the main researcher's consistency and establish how they had come to their conclusions regarding the theme allocation using the literature. This generated a discussion that provided a valuable opportunity to agree upon each theme and title and reflect on their contribution into the core theme and aim of this study (Nash and Sproule, 2012). Following this the main researcher re-analysed each transcript according to the main research purpose and themes to ascertain if any links or associations existed within or between each theme. The main themes and subheadings were then developed to depict the links and networks between components of the core themes using the creation of a conceptual map (Nash, 2008). Figure 6.5 illustrates how data were used to establish categories and themes (Nash et al., 2011). Additional illustrations demonstrating how data were used to establish the remaining themes, similar to Figure 6.5, are available in Appendix 12.

In an attempt to avoid the influence of biased views on the direction of the findings and conclusions and ensure rigour in this research, a qualitative research methodology was used that embraces the concept of trustworthiness in: credibility, transferability, dependability and confirmability (Dorgo, 2009, Nash and Sproule, 2011). This strategy included the use of:

- Member checks: In the interest of trustworthiness or credibility and to ensure accurate interpretations and categorisations of the coaches' meaning, the coaches were given the opportunity to review the researcher's interpretation of the data from their interviews by providing a summary of the coaches' responses. This summary allowed the main researcher the opportunity to verify and ensure accuracy between the researcher's interpretation and coaches' intentions (Dorgo, 2009).

- Peer examination: A colleague with expertise in this research topic commented on the findings and reviewed the interpretation of the themes to confirm the outcomes by deductively analysing a sample of transcriptions (Dorgo, 2009).

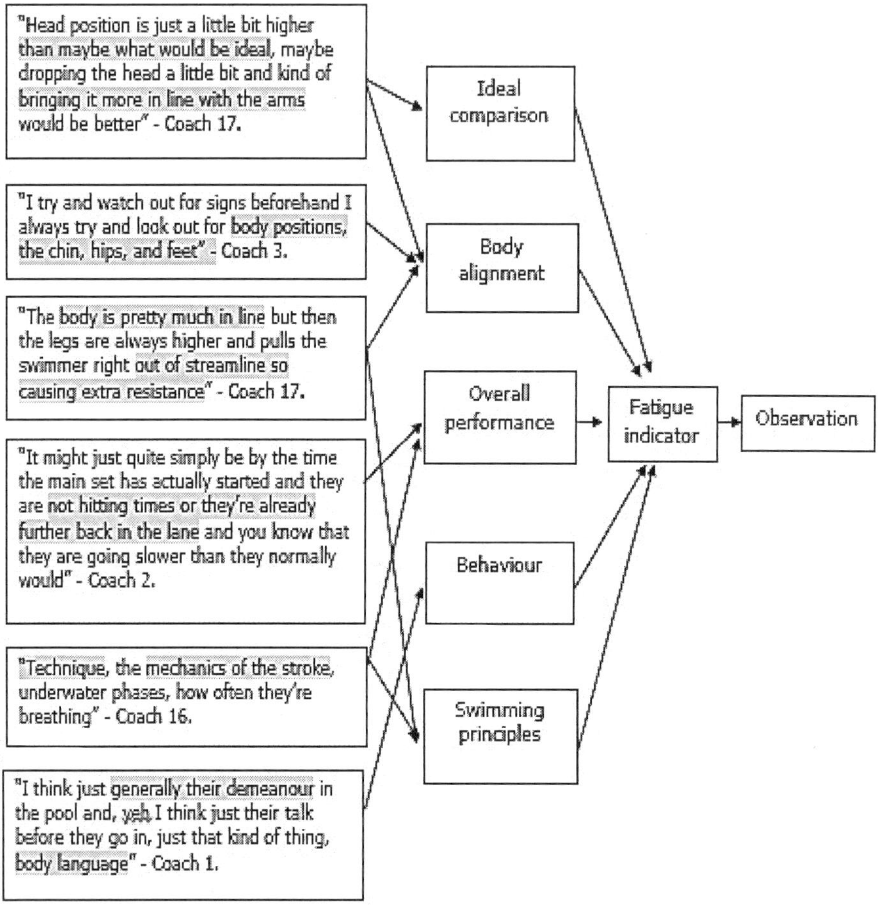

Figure 6.5: An example of the qualitative data analysis (Nash et al., 2011). Those words highlighted are the phrases that link into each category and thus sub-heading.

- Triangulation: The application of triangulation, in terms of the use of qualitative and quantitative methods, member checks, and peer examinations resulted in the use of evidence from multiple data sources. This, as well as the data gleaned from the repeatedly analysed transcription methods, improved the dependability of the findings (Dorgo, 2009). An independent peer audit also improved the researchers consistency and the dependability of the findings (Dorgo, 2009).

These three steps confirmed the analysis of the qualitative data, strengthening its interpretation. Taking these steps is important to ensure the research is rigorous (Nash and Sproule, 2012).

6.3. RESULTS

The results are presented according to the two distinct phases detailed in the methods section: firstly the results of the analysis of coaches' observations of technique and the effect of a video intervention; secondly the coaches' comments and perceptions regarding the observation and identification of fatigue during training are described.

6.3.1. The Analysis of Coaches' Observations of Technique and the Effect of a Video Intervention

6.3.1.1. Baseline test: Session 1

The responses of the coaches were analysed to assess the common stroke features the coaches observed or noted during their observation of the swimmers, as shown in Table 6.2. During the initial baseline test (Session 1) coaches predominantly focused on the stroking parameters of SL and SF, and the leg glide phase. The remaining technique variables were rarely specified and a total of seven variables were not detailed at all during Session 1 (See Table 6.2).

Table 6.2: The number of observations by coaches of each technical variable during each test session. Displ = displacement, L = left, R = right.

	The number of coaches observing each technical variable			
Technique variable	**Session 1**	**Session 2**	**Session 3**	**Total**
Foot displacement	0	0	0	0
Hand displacement	0	2	1	3
Head disp. at breathing	1	8	3	12
Trunk angle during breathing	2	0	0	2
Hand disp. arm out-sweep L	0	0	0	0
Hand disp. arm out-sweep R	0	0	0	0
Knee disp. leg rec L	0	0	1	1
Knee disp. leg rec R	0	0	1	1
Foot disp. leg in-sweep L	0	0	0	0
Foot disp. leg in-sweep R	0	0	0	0

	The number of coaches observing each technical variable			
Technique variable	**Session 1**	**Session 2**	**Session 3**	**Total**
Leg glide phase	4	2	1	7
Average velocity	0	1	0	1
Stroke length	10	9	3	22
Stroke rate	7	6	3	16
25m swim time	0	1	0	1

There was a small difference between the mean percentage of technical variables identified by the experimental group (12.73% ± 11.5) and the control group (9.09% ± 5.68) during Session 1, however this difference was not statistically significant ($p = 0.53$).

6.3.1.2. Post intervention test: Session 2

The most popular technique observed by the coaches following the intervention continued to be SL, followed by SF, as shown in Table 6.2. Following the intervention, the experimental group identified a number of additional variables to those detailed in the initial test session, including: hand displacement, head displacement at breathing, average velocity, and 25m swim time. A smaller number of coaches also identified a change in the leg glide time in Session 2 compared to Session 1. Again a number of variables were not identified by any coaches, including: foot displacement; trunk angle during breathing; hand displacement of the arm during out-sweep; knee displacement of the leg during knee recovery; foot displacement during the leg in-sweep phase.

Overall, the results highlighted an improvement in the proportion of technical variables identified by coaches in the Exp group in Session 2. Only the Exp group improved the proportion of technical variables correctly identified in Session 2, relative to Session 1, with a median of 13.64% to 15.38% and a trivial to small effect size (0.07), however this was not statistically significant ($p = 0.77$). In contrast, the differences in the control group in Session 2, relative to Session 1, showed a decrease in the proportion of technical variables identified, from 9.09% to 7.69%. Although the number of technical variables correctly identified decreased, this was not statistically significant ($p = 0.59$) and also showed a trivial effect size (-0.12). Despite the opposing changes following the intervention/treatment by each group, no statistically significant differences were identified between groups during Session 2 ($p = 0.17$). The percentage median and interquartile scores for the proportion of technical variables correctly identified (a) and the magnitude of change relative to Session 1 as indicated by effect sizes (b) across Sessions 1-3 is shown in Figure 6.6.

(a) Test Session Number

Experimental (exp) group	Median (%)	Inter-quartile range (%)
Session 1	13.64	15.91
Session 2	15.38	7.69
Session 3	4.17	8.33
Control group		
Session 1	9.09	0
Session 2	7.69	5.77
Session 3	4.17	8.33

(b)

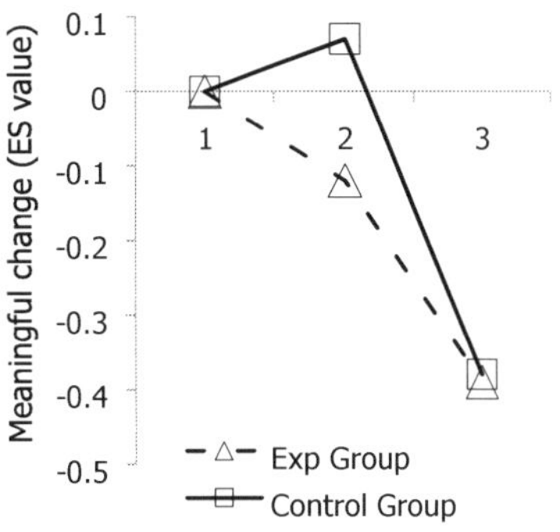

Figure 6.6: (a) Median and inter-quartile percentages of technical variables correctly identified for each group for each session. (b) The magnitude of changes of the percentage of technical variables correctly identified, indicated by effect sizes (ES) relative to Session 1.

6.3.1.3. Retention test: Session 3

In Session 3, the proportion of technical variables the coaches observed or noted during their observation of the swimmers drastically decreased, including the SL and SF (See Table 6.2). Throughout the course of the entire study, three technique variables were not identified by any of the participating coaches, including: foot displacement, hand displacement at the end of the arm out-sweep, and the foot displacement at the end of the leg in-sweep.

Both the control group and experimental group showed a decrease in the proportion of technical variables observed in Session 3, relative to Session 1, which were also not statistically significant for both groups (p = 0.08, p = 0.09, respectively). This decrease had a small-to-moderate meaningful effect for the control group (-0.39) and the experimental group (-0.38). Again, the proportion of technical variables correctly identified was not statistically significantly different between each group in the final test session (p = 1.00).

To assess the retention capacity of the experimental group, the proportion of technical variables identified in Session 3 was also compared to the proportion identified in Session 2. A decrease was shown in the proportion of technical variables correctly identified from Session 2 to Session 3, by both the control group and experimental group, as shown in Figure 6.6. These were also not statistically significant for either group (p = 0.44, p = 0.09, respectively).

6.3.2. Coaches' Perceptions of the Observation and Identification of Fatigue during Training

Each of the coaches was asked the same questions from the interview schedule but due to the use of a semi-structured interview, the depth of the answers varied considerably among the coaches. This resulted in an array of interview durations, generally ranging from five to fifteen minutes. The responses of coaches to the interview were analysed using the guidelines of thematic analysis to allow for the additional responses to be reflected in the results. The analysis of the data and its relevance to the core theme of coaches' observations and identification of fatigue during training resulted in the emergence of three main themes and subcategories. As mentioned in Section 6.2.5.2, illustrations demonstrating how data were used to establish each category and theme, similar to Figure 6.5, are available in Appendix 12. These themes were observation, coaching philosophy, and education. The names of each category were developed to reflect their content and, together with the sub-categories, allowed the creation of a conceptual map (Figure 6.7) to display the main themes, sub-themes, categories and patterns identified from the interview data (Nash, 2008).

6.4. DISCUSSION

The main aim of this study was to investigate whether the use of technical indicators of fatigue in breaststroke swimming can aid competitive swimming coaches' abilities to observe and identify the changes in technique associated with acute fatigue during a training session. The study identified that coaches tended to observe a range of factors when monitoring fatigue, including technical and behavioural factors, and had varying perceptions of fatigue during training. In terms of the observation and identification of specific individual technical changes associated with acute fatigue during a training session, coaches were able to observe and identify changes in a small proportion of technique variables for which 2-D video analysis also showed a change with fatigue, namely: SF, SL, vertical displacement during breathing and the leg glide time. However, the observation and identification of the remaining technical variables was very limited. Following a one hour video intervention, coaches were able to identify a slightly larger, but not statistically significant, proportion of technical variables associated with acute fatigue during a training session. However, the skills developed during this session were not retained.

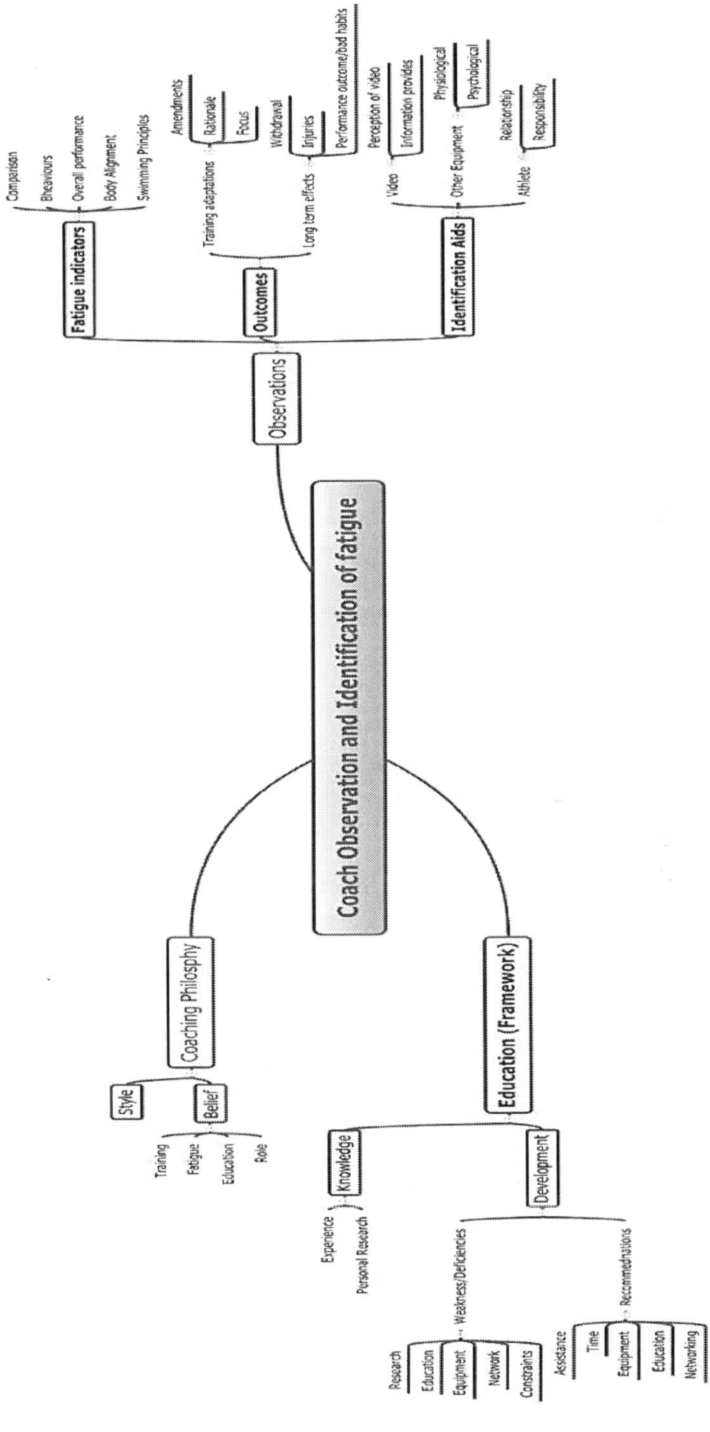

Figure 6.7: A conceptual map of coaches' perceptions on the observation and identification of fatigue.

The discussion section will be presented in a similar format to the results section for ease of comparison. In addition to exploring the conceptual map and inter-relationships to the current literature, the conceptual map will also be linked to the results of the intervention in an effort to better understand coaches' observations of fatigue using technical factors during training.

6.4.1. The Analysis of Coaches' Observations of Technique and the Effect of a Video Intervention

6.4.1.1. Baseline test: the technical indicators of fatigue coaches currently observe

The initial baseline test identified that coaches tend to focus on stroking parameters, specifically SL and SF, when monitoring swimmers for acute fatigue (See Table 6.2). This may be for several reasons; the changes in swim time were strongly related to changes in SF and SL (as identified in Chapter 5) (Conceição et al., 2014); these variables may be more observable due to larger changes, as also shown in Chapter 4; finally, coaches may already perceive these variables as important, (as identified in Chapter 5). As a result there has been a large amount of research regarding the importance of SF and SL in identifying stroke efficiency and effectiveness, which is available both in scientific journals (Conceição et al., 2014, Strzała et al., 2014) and swimming textbooks (Maglischo, 2003). However, research and information pertaining to specific technical variables and the effect of acute fatigue during training is scarce and consequently coaches may not have an in-depth understanding of them (Chapter 5). Since both SL and SF depend upon the individual technical actions to maximise or minimise propulsion and resistance, respectively, coaches may be focusing on these two factors as indications of changes in specific technical components (Conceição et al., 2014, Strzała et al., 2014, Takagi et al., 2004). In addition, Session 1 also identified that coaches attempt to observe a large range of indicators to monitor fatigue during training. These could be categorised under the major technique components of body position, arm and leg actions, breathing and timing (BLABT) throughout the entire swim, including the turn.

Although coaches predominantly identified stroking parameters visually, the remaining technical variables were rarely observed and coaches struggled to identify the remaining technical variables that changed during a high-intensity training session. This was reflected in the low proportion of total variables identified by each coach, per test session (See table 6.2) and may be for several reasons:

The familiarity of the swimmer: Coaches were unfamiliar with the swimmer being observed in the present study. As coaches spend many hours with their athletes, perhaps coaches would be able to identify a larger number of variables with athletes with whom they are familiar.

The familiarity of the observation tool: The coaches participating within this study may have been unaccustomed to the video method used to observe each swimmer, including the use of three camera perspectives simultaneously. The angle of these camera perspectives may also have hindered their observation capacity as they may not have been positioned at the height to mimic exact eye level.

Limitations of above-water observation: Coaches may have not been able to identify the technical changes which may have taken place during an underwater phase of the stroke due to the observation from above-water only.

Limited knowledge on specific technique changes with fatigue: previous research has focused upon stroking parameters when analysing technical changes with fatigue and as a result coaches may be unaware of alternative aspects of technique to monitor.

Further research is needed to determine whether the observation and identification of technical changes associated with acute fatigue during training is related to the familiarity of the swimmer or observation location. The present study was the first to investigate what variables coaches look for when attempting to identify technical changes with fatigue and so future research should consider identifying why these variables cannot be identified or observed.

6.4.1.2. Post intervention test

The overall results of analysis between sessions and groups highlighted that the experimental group were able to identify a slightly larger proportion of technical variables changing with fatigue following the intervention only. As the control group did not improve with repetitive practice, the changes in the experimental group may be deemed to be a result of the intervention and not due to continuous practice. The intervention may have slightly improved the ability to identify the technical factors which change as a result of fatigue for several reasons:

- This method combined verbal and video information, provided by a coach and the Dartfish software's user-friendly graphic and visual images, allowing participants to evaluate performance errors using this information (Guadagnoli et al., 2002).

- Second, Dartfish enabled participants to visually compare incorrect and correct technical performance trials and identify the differences (Thow et al., 2012).
- Third, this provided information on knowledge of results and performancerelated kinematic variables, particularly the technical variables and times, focusing the coaches' attention on limb actions and external effects of actions. Some research has shown that focusing individuals' attention on the external effects of actions greatly improves skill development (Wulf and Prinz, 2001).

Although the proportion of variables identified increased slightly in Session 2 for the experimental group, the variation in the proportion of technical variables correctly identified, and lack of a statistically significant improvement, could be for a number of reasons: firstly, as three different swimmers were observed by the coaches (see Figure 6.2), coaches may have found difficulties in observing one swimmer compared to another due to individual differences in technique. However, due to the novelty of the study and need to standardise the observations amongst coaches, this was also necessary. Secondly, due to the data not conforming to the assumptions of parametric tests, non-parametric statistics were used which could have reduced the statistical power, resulting in the potential for increased type 1 and 2 errors, and may have also reduced the capacity to identify the small changes in the proportion of technique variables identified by the coaches. Thirdly, the variations in each individual's knowledge level and skill (Hodges and Franks, 2002, Leas and Chi, 1993, Moreno et al., 2006); and differing individual adaptation rates to feedback, could have influenced each coach's capacity to observe technical markers related to fatigue (Guadagnoli et al., 2002). This was emphasised by the limited consistency of changes relative to Session 1 following the intervention, with some coaches improving the proportion of variables they observed, some decreasing the proportion of variables they observed and some not changing at all. Although the present study was designed to enable assessment of the effect of intervention on learning to identify technical fatigue during swimming, and a nofeedback option was used, the various formats of feedback were not assessed. For example, the amount, type, or duration of feedback (Hodges and Franks, 2002). The present study has played an important role in initiating research into this topic and establishing a feedback method which elicits some limited improvement in the identification of technical variables after only a one hour training session. Future research should not only expand the training duration of the intervention but also apply the information and observational practice in a poolside environment.

6.4.1.3. Retention test: Session 3

In the present study, a retention test was used to reduce the possibility of misinterpreting improved performance (Magill, 2004). Magill (2004) stated that four weeks was a sufficient period of time in order to ascertain if newly learnt skills could be retained or not. Unfortunately, this was not sufficient in the present study to elicit consistent or improved performance. Instead, a month without further information or feedback following Session 2 resulted in a reduction in performance of the experimental group, indicating that any skills developed were not retained. This may be a result of: the use of a different intervention protocol consisting of only one session; the lack of practice sessions; and the use of a different graphic video software. The control group, who received no information, only showed deteriorating performance from Session 1 to Session 3. This could be explained by the effect of 'observer drift', or the observers' 'tendency to change' the factors they observe or interpret the technique differently over time (Mets et al., 2003). The deterioration in performance by both groups in Session 3 could again be due to the level of swimmer being observed and their technical performance. The largely reduced proportion of variables observed by all coaches for Swimmer 3 could imply that the coaches had difficulties observing this swimmer in comparison to the other two swimmers in Session 1 and 2. Elite level swimmers were used in the present study due to the likelihood that stroke characteristics and stroke patterns would be well established and consistent (Pyne et al., 2004) and any changes in technique could be related to acute fatigue from a high-intensity session, as was also found in Chapter 4.

From these results, it can be noted that 'learning' did not occur in the experimental group due to the fact that the skill showed improvement but was not 'consistent or persistent' in its use by the coaches (Magill, 2004). Rather, the evidence suggests that the video- and computer-based intervention could acutely improve performance. Using various formats of feedback could influence the retention capacity of the coaches and thus explain why the information was not retained. However, more work is necessary to gain insights into the most effective combination of feedback approaches (Hodges and Franks, 2002). Further research is needed to determine the optimum environment and feedback methods using video to develop this skill amongst competitive swimming coaches.

6.4.2. Coaches' Perceptions of the Observation and Identification of Fatigue during Training

The interviewed coaches displayed a number of similar characteristics: observation, coaching philosophy and education. Each of these characteristics contains a variety of subcategories and inter-relationships which emerged from the interviews and allowed the development of the conceptual map (Figure 6.7) (Nash, 2008). These factors will now be discussed, with consideration of the current literature and results from the intervention.

6.4.2.1. Coach Observations

The technique variables coaches identified during their observation of the video ranged throughout the entire body and stroke, including areas such as body position, arm, leg, and timing actions. These observations could be further divided into behaviour, overall performance, body alignment, and swimming principles. Behaviour included the personality and attitude of the swimmer, with Coach 1 stating:

"I think just generally their demeanour in the pool and, yeh I think just their talk before they go in, just that kind of thing, body language."

This implied that coaches not only observed aspects in the water but also the athlete themselves. Mood changes have been shown to be a large sign of fatigue (Ament and Verkerke, 2009) with variations in mood being associated with higher or lower training intensities. Body alignment was specifically linked to the body in relation to the water level and limb actions:

"I try and watch out for signs beforehand, I always try and look out for body positions, chin, hips, feet." - Coach 3

Despite the range of fatigue indicators observed by the coaches, the technical focus was upon the SL, SF, vertical height during breathing and leg glide duration. The remaining variables were only identified once or twice throughout the entire study. In addition, certain variables were falsely identified as changing or changing in the wrong direction, including SF and hand depth. Although coaches were informed of the changes that did occur following the intervention and retention test (depending on the group allocation), no discussion took place as to why coaches observed certain variables but did not observe others. Some potential reasons for this may be that: coaches had difficulties identifying changes in technique using visual observation due to the high level of the swimmers or speed of movements; coaches were not observing those specific parts of the technique; or coaches could not observe these changes due to the fact they took place during underwater actions which are difficult to observe from above the water (Leas and Chi, 1993). Future research could enquire about coaches' perceptions of their observations.

In addition, several coaches were shown to not only identify technical factors but also relate these to other aspects of performance. Some coaches compared the actions to elite models, stating comments such as 'compared to the ideal' or describing what technical actions should be. Other coaches related it to the principles of swimming, including propulsion or resistance, with comments such as 'his kick lost pretty much all propulsion'. Finally some coaches just stated the overall performance of the athlete during the session was a good indicator of their fatigued state:

"*Technique, mechanics of the stroke, underwater phases, how often they're breathing.*" - Coach 16

From this it can be gleaned that coaches are not simply observing one factor when monitoring fatigue but a multitude of factors related to fatigue. The monitoring of a multitude of factors is one role of the coach that has been described many times (Mets et al., 2003, Moreno et al., 2006) and was emphasised in the current study as coaches also commented upon the observation of the outcomes of fatigue, specifically the short-term training adaptations and long-term effects or consequences of unattended fatigue. The training adaptations were summarised into the adaptations coaches would make to ensure their athletes did not become too fatigued, Coach 13 stated:

"*So I would give more rest so they can recover into some reps or give them some active recovery in sets compared to everyone else or make them rest, and give them more sessions off and resting.*"

The responses of the coaches again varied, but the predominant variable coaches stated they changed was the rest duration within a set. Although the stress induced during training is a vital aspect of athlete development in order to promote adaptations, recovery plays an important role in this process (Meeusen et al., 2013, Smith, 2003). Their focus and rationale for making these changes was very much dependent upon the individual coach's purpose or goal:

"*It would depend what the purpose of the session was.*" - Coach 2

These responses were similar to those provided in the previous chapter as well. The opposing rationale of the coach attempting to maximise the outcome of each training session was to prevent the long-term consequences that fatigue could impose upon their athletes. These included injury and withdrawal from the sport, but mainly the prevention of bad technique. Coach 1 stated:

"I think what you do in training you do in a race. Because you are stressed you will just fall back and do what you do in training. If you do sloppy turns in training you are not very likely to do good turns in a race."

This is also in accordance with the processes of motor learning and the importance of practice in the automation of motor skills (Bonacci et al., 2009, Williams and Hodges, 2005). However, it should be noted that 'sloppy' turns may not be indicative of fatigue alone and could simply be swimmers who are not motivated to train. Despite this, the concern of performing technical actions incorrectly during practice in terms of the automation of bad technique into bad habits which could influence race performance was apparent amongst all the coaches. Coach 5 stated:

"If you practice perfection you'll get perfection, if you practice mistakes you'll get perfect mistakes."

The identification of this range of factors indicates that coaches consider a wide range of factors worth observing in the identification of fatigue, from the acute aspects viewed during a single stroke cycle to those viewed over weeks to months of training. This is important as it represents the dearth of effects which fatigue can cause (Ament and Verkerke, 2009).

To aid this process, coaches stated the use of other tools and feedback aids were essential to allow the coach to monitor and manage each swimmer. Often the use of physiological tools, such as heart rate monitoring to assess the training load, or psychological tools to assess the athlete themselves were stated as vital. In the previous chapter, the main tool used every training session was the coach's own visual observation, followed very closely by the stopwatch and above-water video camera. The present study may answer why this was the case as the coaches in the present study stated mixed perceptions regarding the use of video. Some coaches stated it was an advantage due to its ability to analyse aspects in depth:

"There is obviously advantages of watching from video... you can take it back, you can replay it, you can analyse it you can look at different things, you can concentrate on one thing at a time, play it again and look at something else and what have you." - Coach 2

Despite these advantages, some coaches still stated that they preferred observing without the aid of video and stated that its application could be a disadvantage due to the time it takes to use:

"If you are doing quality then it uses up more water time and if you are spending more water time you have to balance out whether the quality you are doing by being out of the water more is going to have

a significant impact on what happens in the water and I can't really comment on that because I haven't done enough of it to assess whether it's successful or not. I suspect it probably would be but it's that moment where, how, how often do you then video analyse swimmers." - Coach 8

Despite the clear interest in observation as a topic, a common theme amongst all coaches was the limitation of observing and managing fatigue as a result of the numerous roles coaching involves and applying this to each and every individual swimmer (Nash and Sproule, 2009). Coach 19 stated:

"When you think about having thirteen swimmers they must all be affected differently and you have to identify which part of that applies to that swimmer and how you can help that swimmer either by cutting down, changing technique or the way that you present your set and that's a difficult task."

This was also noted in terms of the other jobs coaches have to contend with on poolside:

"The problem being that poolside there is all sorts of other things going on so invariably somebody is asking you something or something else is happening it is very hard to focus just on what you're doing." - Coach15

Finally, the athlete was considered a key factor in the identification of fatigue. This was highly related to the coach's relationship with the athlete:

"I think it is sort of trying to make sure you have that proper relationship with the swimmer too so that you trust them and they trust you to work together." - Coach 3

Part of this is reliance on the athlete to provide information:

"I rely on my swimmers a lot to feedback that to me." - Coach 8

Some coaches also noted that this was vital for the development of the athlete and their own responsibility in their own involvement:

"And it gives them sort of the responsibility of trying to maintain their form during the swimming as they go through the set." - Coach 3

This introduces the concept as to why coaches are interested in tools which are fast, efficient and effective at providing information to large groups of athletes (Fleming et al., 2010, Mets et al., 2003). This is largely related to their day-to-day experiences in the role of a coach. Although there are a lot of common comments related to those responses found in the questionnaire in Chapter 5, the observations and methods coaches used still varied greatly depending upon the coach. In addition, it was also similar in that although the coaches stated a range of factors, these all varied greatly between coaches.

Apart from the stroking parameters, the remaining variables were mentioned very independently. There was also a difference in the way the coaches described these variables, with some only stating what they observed in terms of the name of the variable and what changed about it. Others were shown to elaborate on this, linking variables and statements together and attempting to explain why a variable changed a certain way or not. As an example, Coach 6 stated:

"As the swimmer gets further into the 100, their arms extend across the centre line to the left hand side so there is obviously a little bit of tension appearing in the upper back at that point which then shows through the phase changing durations."

According to Leas and Chi (1993) and Moreno et al. (2006), this demonstrates an ability to identify with the causes of stroke errors and is often associated with a more expert coaching level (Leas and Chi, 1993). However, in the present study a mixture of responses was found, with both more novice or elite coaches stating both short or long and linking comments. Reasons for this may have included: the novelty of this topic, the education format, the intervention method, the duration of the video clip and perspectives observed from. Despite research stating some unique differences between elite and novice coaches, this was not in the scope of this thesis. Further research is needed which specifically investigates coaches knowledge and application of this topic, with specific interest in differences between elite and novice coaches.

6.4.2.2. Coaching philosophy

The approach to fatigue observation and identification of fatigue by the coaches varied greatly. Not just regarding which processes or methods to use but actually about the coaches' beliefs on the topics itself. These varied greatly amongst all the coaches in terms of their perception of their role as a coach, education, training, and fatigue. Coaches often perceived their role to focus on the improvement of the athlete and to achieve this in the best and most efficient way possible. Ensuring access to all the knowledge and information coaches can to educate themselves was a key aspect of achieving this:

"The more proficient we are at knowing which improvement happens the easier it is to communicate the feedback with them." - Coach 6

This knowledge may influence how coaches train their athletes with some perceiving the quality of training as a vital aspect, Coach 8 stated:

"So to me it is always based on the quality of what they do with an eye on how fast they are doing it, as opposed to how fast they do it with

an eye on the quality so I prefer to look at the quality of the technique than time."

And others noting that quality is not everything, with Coach 13 stating:

"Well it depends how quality the quality is and also you can get high quality/quantity as well. I mean I would rather have someone who can swim one 400 fast than say a million 400's fast at average speed. Because there is fit to race and there is fit to train so long as someone's is fit to race its fine."

The debate between quality training or quantity training is on-going in swimming as coaches strive to help their athletes to attain their full potential by maximising the efficiency of their training (Arroyo-Toledo et al., 2013, Richmonda et al., 2015). Unfortunately, the perception of fatigue and recovery in this process also differed widely between coaches. Some coaches perceived fatigue as a necessity of training:

"I find that during training sessions it is important to during certain sessions fatigue the body so that swimmers are aware of what happens to their stroke, what biomechanically what isn't as good or what isn't maintained as well." - Coach 17

Others perceived it as a limitation which needs to be carefully monitored:

"Yes it becomes meaningless training when there is tiredness creeping in they are going through the motions but psychologically its, it is defunct, you know it's a negative and when negativity comes into swimming you should really move away from it." - Coach 5

This was in accordance with Chapter 5 which also found very varying opinions of fatigue. Beliefs about these factors all influence how coaches implement their training plans and decisions and may also have biased what coaches chose to observe. Since the beliefs and styles of coaches varied amongst these coaches, regardless of experience, it can be insinuated that coaches are doing very different things when it comes to observing and managing fatigue during training.

6.4.2.3. Coach Education

The variations in coaches' observations and perceptions of the areas of fatigue and training could be a result of coaches' education and where they source their initial knowledge. As in the previous chapter, it was again noted that coaches sourced their knowledge of this topic from experience. Some coaches even stated that they collected their knowledge from personal research which the coach undertook in their own time:

"I do find bits and bobs on the internet." - Coach 3

However, the collection of knowledge from the internet alone may be risky as not all information provided from the internet is based on evidence or peer-reviewed research (Bartlett, 2007). The need to search for information from the internet was stated to be due to a lack of information provided in coach education courses:

"*But you know here is me on my level 3 and there's been no mention there has been very little in fact I can't recall anything come in on levels 1, 2 or 3 into fatigue.*" - Coach 2

This is in accordance with other literature that states coaches are not obtaining the relevant information from coach education courses (Nash and Sproule, 2009). Coaching courses rely heavily on the coach educators who deliver the courses to be both knowledgeable and able to present the information effectively, both theoretically and practically (Nash and Sproule, 2012). However, Coach 17 even stated that they felt they had more knowledge in the sport science section than the instructor:

"*I had to tell them all the answers to the group and pretty much give a lecture to the group for each question.*"

If coaches cannot obtain their knowledge from standard sources and choose to do personal research, this means they rely on the information they are finding and obtaining being correct which is not always the case. This is vital as it then influences what a coach observes and thus what they feedback to the athlete. If incorrect this lowers the quality of feedback the coach is providing (Moreno et al., 2006). This was seen to be common, regardless of the coaching experience or level in the present study despite the fact coaching at the beginner level is very different from coaching at the more elite levels (Nash and Sproule, 2011). However, according to this group of coaches, even CPD courses for coaches are limited and lack information on the topic of fatigue or the use of video during training for all ages or level of swimmer:

"*A lot of the CPD's we do are either learn to swim or like performance there is not that many in the middle for age groups not that I've seen anyway.*" - Coach 11

This was noted in the present study as coaches stated the limits of present education courses, regardless of their experience or education level, stating there was not enough education courses on coaching itself and the focus is more on improving the athlete:

"*No like, in terms of teaching the coaches there isn't very much.*" - Coach 17

To attempt to improve this, the coaches in the present study repeatedly requested better education on the topic of fatigue during training and the use of video for all performance levels and for coaches and swimmers. This is an essential start in the development of coach education programs, with current literature suggesting that the inclusion of coaches in this process will enable the production of more effective courses (Nash and Sproule, 2012). By including the coaches in the development of the present research, this data has begun this process with the coaches requesting further information and training sessions on the use and application of video to monitor athletes during training. To continue the application of this research, this information could be disseminated to institutes of sport and coaches throughout Britain using workshops, courses and access to this data. This is important to ensure coaches are kept up to date with current research and findings and will be elaborated upon in Chapter 7.

6.5. CONCLUSION

In conclusion, this study provides evidence that coaches tend to observe a variety of indicators of fatigue, ranging from technical actions throughout the entire body to behavioural actions of the swimmer. Although there were a number of similarities amongst the coaches which enabled the grouping of categories and themes regarding this topic, the individuality and personal approach to dealing with this topic in training was a large factor. This highlights the individuality of the coaching methods as well as the lack of guidance on this topic. The knowledge coaches had obtained linked to their perception of this topic and thus their observations and approach to its identification and management.

Coaches were able to identify stroking parameters and leg glide duration consistently during their observations; however, the remaining technical variables were rarely identified. The use of 2-D video and analysis software training showed some potential to improving the ability of coaches to identify the technical changes associated with fatigue during training after only a single one hour training session. These changes were not statistically significant and skills were not retained, suggesting that the intervention protocol could be further developed to improve the retention of the knowledge and observational skills. Another clear outcome from this study was the lack of education on this topic or the use of video in the present education courses and continued professional development for coaches.

Other limitations coaches commonly stated were related to the lack of access to equipment, the constraints of coaching itself in terms of time and means to implement any education they receive. Finally, although coaches often obtain knowledge through experience, a lack of networking amongst coaches was identified and it appears to be a common notion that coaches are very individual and isolated in their methods, and reluctant to share information, with Coach 6 stating:

"Lots of people use it, lots of people video lots of people but there is not really, certainly not met many people that will have that conversation about where they are making changes, why they are making changes and how they are making change."

Future research may therefore involve assessing coaches' observations and identification methods in an actual training environment, investigating differences between novice and elite coaches and assessing other video and computer interventions to determine the most effective method.

Coaches attempted to aid the development of future research in the present study by describing factors they stated were needed to improve the education process and the observation, identification and management of fatigue by coaches in the future. This included further education, better collaborations with varying levels of coaches, access to more equipment, time and assistance to implement such plans and research which did not focus on elite athletes and coaches only. This information and research is needed to enhance coaches' knowledge and ensure they are obtaining the correct information through effective coaching education program and methods. In an effort to further coaches' education of this topic, the data from this study could be researched further and developed into an aspect of professional development.

Chapter 6: Summary

What was already known about this topic?

- Coaches predominantly use visual observation, stopwatches and abovewater video analysis to monitor their athletes during training.
- Technical changes which occur during high-intensity training are measurable using 2-D video analysis.
- Video feedback improves athlete performance but very little research has been on coach performance or its use during training in the identification of fatigue.
- Education on fatigue monitoring during training is limited.

What new information does this chapter provide?

- Coaches tend to observe a range of factors when monitoring their athletes during training; including: athlete behaviour, swimming principles, overall performance, body alignment, training performance and the presence of long-term effects.

- Coaches were unable to consistently observe and identify a number of the changes in technique which occurred during a high-intensity training session. The two predominant variables continually identified correctly were the stroke rate and stroke length.
- A one hour video training intervention was shown to slightly improve coaches' abilities to identify the technical changes which occurred during a high-intensity training session, however this was not statistically significant and only a small effect was shown.
- Coaches' perceptions of fatigue and training vary widely between coaches, regardless of experience.
- Coaches feel there is a lack of use of video in the current coach education and CPD system in swimming. Coaches would also like more training on the topic of fatigue and the use of 2-D video and video analysis.

7

General Discussion and Conclusion

7.1. OVERVIEW OF FINDINGS

The purpose of this thesis was to explore the implication of video analysis methods available to swimming coaches to aid the monitoring of fatigue during training. This thesis has identified that 2-D video analysis is an effective and useful tool which can monitor fourteen acute technical changes and swim time and has the potential to aid and develop coaches' observations of technique during training. The use of diverse and innovative approaches helped to identify the potential role and impact of 2-D video methods and the barriers to its use by swimming coaches during training sessions. This was vital due to the importance of technical actions for successful performance in swimming and due to the negative influence fatigue can have on these actions during phases of high-intensity training.

Despite technological advancements in video equipment and exploration of the topic of fatigue, research into the application of video for coaches to monitor the effects of fatigue on technique during training has not taken place. Hence, a gap was shown to be evident between the theory and use of 2-D video analysis in research and its application in practical coaching environments. This has resulted in an inconsistency between the methods and approaches used by coaches and sport scientists to monitor athletes' performance during training. Due to the scarcity of previous research on this topic, this thesis focused on addressing the inconsistency between the biomechanical analyses and coaching application of 2-D video methods by challenging the traditional methods of monitoring fatigue and feedback in swimming during training.

7.1.1. Study One

Study one was conducted to investigate whether 2-D video analysis methods can enable accurate, precise and reliable measurements of angle, distance and timing variables of breaststroke swimming. To achieve this, thirty-four technical variables and swim time were measured using Dartfish video analysis software (Dartfish Ltd, Fribourg, Switzerland) and compared to the smallest worthwhile change. By calculating and comparing these technique variables, those which could be measured objectively, with accuracy and precision, were determined.

A total of fourteen kinematic technical variables and swim time (out of a total thirty-five dependent variables) were found to be valid and reliable, both between-subjects and across trials, within-subjects. Certain technical variables had large errors, inaccurate estimations of angles and distance values and unreliable measurements. The largest errors were for angle measurements taken when the limb was out-of-plane of the camera, and specifically from a front-view camera perspective. This is in accordance with previous literature and relates to the advice of biomechanics literature regarding issues with measurements taken out-of-plane of the camera (Bartlett, 2007). These findings suggested that 2-D video analysis can be used to measure certain technique variables and has the potential to be used in a training environment; however, coaches should monitor certain variables with caution due to large errors caused by the measurements being taken out-of-plane of the camera.

Given the role of coaches in detecting changes in variables that are related to fatigue, it is essential that coaches know which variables to observe. If a coach's feedback based on their observations and interpretations is incorrect, the swimmer's performance could be adversely affected. This is particularly true in technique analysis due to the role technique plays in overall swimming performance. Thus, it is recommended that coaches avoid measurements which were not shown to be valid, reliable or precise, such as those from a front-view camera perspective.

This was the first study to quantify breaststroke technique measures in the form of angle, distance and time variables using 2-D video analysis software and to assess whether such methods can produce data of sufficient quality to monitor swimmers during training. This advances current literature by overcoming some of the difficulties of video analysis and challenging the application of 2-D video analysis into a water sporting context. This chapter was used to report the stages of technical development that have been undertaken in an attempt to address the primary research questions.

7.1.2. Study Two

Study two determined the use of the set of technical markers identified in Study one to investigate technical changes due to fatigue during a high-intensity training session. Due to the lack of literature on this topic in the context of training, understanding of these changes is limited. The second aim of this study was to assess whether these changes were common among elite level national swimmers or specific to individual swimmers. Using a specifically designed swim set to induce a fatigued state representative of highintensity conditions, the fourteen technique variables and swim time (a total of fifteen dependent variables), identified in study one, were analysed from video recordings for each swimmer. Not only did this identify that the technique variables were measureable during a training-like set using 2-D video analysis methods, but also that the set was sufficient to induce acute changes in the performance of these variables. Although this was the first study to analyse the acute effects of fatigue during training, certain changes, specifically the stroking parameters, identified in the present study were comparable to findings from previous research analysing the acute effects of fatigue during race performances in front-crawl (Alberty et al., 2009), and a single study on 200m breaststroke (Conceição et al., 2014). This emphasises the influence that fatigue can have upon breaststroke technical performance, including during training, and suggests that the acute technical changes observed during a race can also occur during a single highintensity set. These findings are important as it is during training that athletes develop and refine technical actions to prepare for competitive performance (Bonacci et al., 2009). Although only acute fatigue effects from a single high-intensity set were analysed in Study two, swimmers often complete block phases of high-intensity training which would involve multiple sessions. If swimmers are experiencing acute fatigue repetitively during high-intensity blocks of training, this could develop sub-optimal technique patterns which become automatic due to repetitive practice for long durations in a fatigued state (Richmonda et al., 2015). This has implications for athletes' capacity to cope with acute fatigue in training and a race situation and future research could investigate the long-term implications of such technical changes in terms of injuries or muscular imbalances.

The swimmers in the current study were able to continue swimming at a decreased pace despite experiencing fatigue and apparent changes in technique. These responses may be reflective of the gradual changes with fatigue from the beginning of the exercise which highlight the extent to which the body is initially able to manage and cope with the exercise and these effects before reaching a state of failure, as suggested by Kelly (2007). This study highlighted gradual individual responses in the ability to maintain initial swimming technique, differences between swimmers and technique variables in terms of when the technical changes initiated, and the format of those changes. This underlines the individualistic nature of swim performance and fatigue (Ament and Verkerke, 2009, Maglischo, 2003).

Both swimming performance and fatigue can be influenced by a number of factors which can differ between individuals including: physical capacity, variations in technical style and the association between efficiency and energy costs during swimming (Ament and Verkerke, 2009, Barbosa et al., 2008, Maglischo, 2003). These gradual, individual changes outside of the established normal technique range suggest that there is a process of adapting to cope with fatigue during high-intensity training and that this process differs between swimmers. This is in accordance with the literature pertaining to fatigue which states that fatigue can be individual and progressive from the initiation of exercise, with accumulating effects when insufficient recovery is provided and the potential to lead to overreaching or overtraining (Ament and Verkerke, 2009, Meeusen et al., 2013). Although individual differences existed, some commonalities were also found in the direction of change of the variables which showed the largest statistically significant changes (leg glide duration, swim time, SF, SL, and average velocity) amongst the swimmers. This is important as it highlights several technical variables which coaches could use to monitor the general performance of a group of swimmers and assess the overall performance of a squad during a high-intensity training session. The changes in these parameters suggest that fatigue may be influencing the stroke efficiency and muscle activation, both of which are important to maximise training effectiveness and maintain optimal speed as required in a race. This implies that coaches could monitor their swimmers using variables such as stroking parameters; however, coaches need to be aware of the differences between individual swimmers.

This was the first study to assess the effects of acute fatigue on specific breaststroke technical factors during training conditions using 2-D video analysis methods. It is also one of the first studies to challenge traditional fatigue analysis tools and coaching feedback methods by validating the use of the 2-D video methods in the measurement of technical variables and their changes as a result of fatigue. This implies the previously established variables and 2-D video analysis methods could potentially be used by coaches during training to monitor the technical performance of their athlete without necessarily requiring exposure to time-consuming 3-D video analysis methods. These findings add to the current perceptions of fatigue effects during training and provide an alternative and innovative method to monitor technical measures and individual variations of fatigue during high-intensity training.

7.1.3. Study Three

Understanding coaches' current perceptions and practices of monitoring fatigue during training is essential to understand whether the use of 2-D analysis methods and the previously established technical variables has implications for coaching practice. Thus, coaches' current practices and knowledge regarding fatigue during training were explored in Study three.

Coaches' responses to a questionnaire revealed that up to 98% of the coaches considered fatigue, its effect and management, important in the development of their swimmers. Statistically significant associations were identified between coaches' qualification levels and their familiarity with certain mechanisms of fatigue, additional factors which can influence fatigue, and use of certain pieces of equipment to monitor fatigue ($p < 0.05$). This indicated differences in the perceived knowledge of fatigue and equipment used during training with coaches of differing qualification levels. However, in addition to this, there was a lack of consistency among coaches in terms of their perceptions of fatigue during training and the methods they stated they used to manage fatigue. This highlights a range of understanding of fatigue amongst this group of coaches and suggested that their education of this topic may not have been complete. This was emphasised by some responses from coaches which stated that education on the topic of fatigue in the coaching programme is limited. Further education of this topic is required due to the important role it plays in many aspects of performance and preparation. This suggests that the current education programme may need to be revised to include further data on fatigue and its role during training, as well as continued CPD courses to keep coaches up to date with any developments in research.

A large number of coaches reported that they monitored and adapted training sessions based on the individual's ability to cope with the intensity of training and rest durations. Coaches stated they utilised a range of methods to monitor fatigue during training. Although technology has developed, coaches stated they were continuing to use traditional methods to monitor their athletes which are quick and reliable, specifically stopwatches, visual observation and self-questionnaires, regardless of coaching experience. The most popular video analysis method was above-water cameras, which with recent developments can now include iPads, iPhones, cameras and tablets. The reasons identified for this was the lack of accessibility, time and understanding and the cost of other pieces of equipment. This suggests that although a wealth of methods are currently available to monitor fatigue, coaches perceive that they are not useable within the training environment. These results are in accordance with one similar study by Taylor et al. (2012) which assessed the monitoring of fatigue in individuals involved in assessing performance in a range of sports, including a small number of swimming coaches. Although this study identified that coaches are utilising above and below-water cameras to monitor their swimmers, it is still unclear whether these method are being used effectively.

This is the first study to have investigated coaches' perceptions of fatigue during training in any sport and the first to analyse the methods they use to monitor fatigue in competitive swimming. This adds to current literature by finding a starting point for further research and identifying where potential gaps or limitations lie in current knowledge or practice.

7.1.4. Study Four

To assess what coaches observe and whether they can identify the technical changes with fatigue, two groups of competitive swimming coaches observed a series of abovewater videos of three individual swimmers and were asked to identify any technical factors which changed as a result of fatigue. Coaches tended to observe a range of factors when monitoring their athletes during training; including: athlete behaviour, swimming principles, overall performance, body alignment, training performance and the presence of long-term effects. This is in accordance with literature on fatigue and its effects, as well as noted effects on swimming performance (Alberty et al., 2009, Ament and Verkerke, 2009, Taylor et al., 2012). This implies that these coaches have an understanding of fatigue and the potential negative effects it can induce in their swimmers; however, it also suggests that coaches are not fully informed regarding the specific effects of fatigue in swimming. This information is needed in order to enable coaches to observe, and thus monitor, the effects of fatigue as much as possible. This has potential impact on coaches' observations of fatigue during training as key factors may be missed while monitoring.

In terms of the previously established technical variables, coaches were effective at observing and identifying changes in four of the previously measured fifteen dependent variables; specifically SL, SF, leg glide duration and vertical height during breathing. This is important as it indicates that coaches could visually identify those technical variables that were the most valid and reliable and which displayed the largest changes during a training set, without the use of video. However, there were some observations of these stroking parameters which were judged incorrectly by the coaches. This suggests that although observable on camera, coaches' visual observations may not be reliable nor consistent. Coaches were unable to consistently observe and identify the remaining technical changes in technique which occurred during a high-intensity training session, including the swim speed, despite coaches stating they observed the entire body. This suggests that coaches can only observe certain variables from above the water and this could be for two reasons: firstly, coaches cannot visually observe them from above the water or poolside due to water viscosity and bubbles, secondly coaches do not consider the technical factors worth observing when looking at fatigue. These two outcomes suggest that the feedback regarding the technical changes which occur due to fatigue could be limited if coaches only use visual observation. Use of video may be a useful supplement or alternative to visual observation to overcome this limitation.

To assess whether coaches could improve their visual observation through an education intervention, one group of competitive swimming coaches underwent an education intervention using Dartfish software (Dartfish Ltd, Fribourg, Switzerland) while a second group of competitive swimming coaches acted as a control group and received no feedback. Following a one hour video training intervention, those coaches who underwent the one hour video intervention slightly improved their ability to identify the technical changes which occurred during a high-intensity training session, however this was not statistically significant ($p > 0.05$), nor retained, and only a small effect was shown.

The effect of video feedback on performance is well established in the literature (Bertram et al., 2007, Wilson, 2008) but the present study is the first to have assessed its effect on improving coaches' observation skills of technical changes with fatigue. This suggests that 2-D video and fatigue education programmes have the potential to improve coaches' observation skills. However, the lack of statistically significant changes and retention capacities proposes that establishing the optimal learning methods, in terms of feedback duration and styles, may help to maximise coaches' observations of fatigue and the impact of 2-D video. This is important to maximise coaches' observations and identification of fatigue to aid the process of monitoring training load, optimising the effectiveness of training and minimising negative effects of fatigue.

Despite the growing use of video by coaches and its development in swimming, the coaches in the present study stated that they continue to face barriers in the use of video during training. This is also in accordance with previous literature regarding the barriers coaches face using certain pieces of technology and equipment (Taylor et al., 2012, Wang and Parameswaran, 2004). Coaches stated that they felt the current coach education system lacked sufficient information on fatigue during training or the use of video, and requested further courses and data on these topics. This study adds to the current literature identifying that the methods coaches are using are capable of identifying technical changes with fatigue; however, the use of 2-D video may enhance this ability by providing additional key information. This is important to maximise the potential of feedback coaches can provide to their athletes.

7.1.5. Summary

According to Bishop (2008a; p. 253), '*sport science can be thought of as a scientific process used to guide sport performance*'. Despite this, the application of sport science research to practice is perceived as being poor (Bishop, 2008a, Martindale and Nash, 2013). This is especially true in biomechanics, where there are issues of application, overuse of quantitative methods and a need for research on the effectiveness of implementing biomechanics into coaching (Knudson, 2007). The present research reinforces this issue through the identification of barriers perceived by coaches in the use methods, including 2-D video analysis, to monitor fatigue during training. The barriers identified by the coaches were in agreement with previous literature which highlighted that current research has failed to study problems relevant to coaches or use equipment which can be applied within a practical environment (Kilic and Ince, 2015, Knudson et al., 2014). The information sport science can offer is deemed a substantial part of the 'knowledge base' needed by coaches and thus it is crucial that research and application can adapt to ensure coaches can receive this vital information (Bishop, 2008a, Martindale and Nash, 2013). This research challenged the current methods and processes of sport biomechanics, video technology and coaching feedback, in an effort to begin addressing the divergence that currently exists between coaches and sport scientists, using original methods. The results indicate that 2-D video analysis is an effective and useful tool, which has practical applications in monitoring fatigue during a training session and developing coaches' identification and management of fatigue during training through education programmes.

7.2. IMPLICATIONS OF FINDINGS

If, as the current studies suggest, coaches and sport scientists can both utilise 2-D video technology in the observation and monitoring of fatigue during training, then this could have a wide range of implications for a number of individuals involved in sport performance, both directly and indirectly. This will now be discussed.

7.2.1. Coaches

This research identified that monitoring fatigue is an important aspect of the training process and coaches can visually identify the largest changes in technique due to acute fatigue (SL and SR). However, the study also identified that, through the use of 2-D video analysis, additional acute changes in breaststroke technique during training could be measured and monitored. By missing these variables, coaches could overlook key technical changes which could result in coaches obtaining or providing incorrect feedback that may be detrimental rather than beneficial to the athlete. This suggests that by utilising the 2-D video analysis methods during training, which have been proven as valid and reliable, the feedback athletes can receive on a daily basis from coaches may be enhanced. In addition, it provides a basis upon which to begin assessing other tools coaches are using to provide feedback to their athletes during training. Information regarding the technical changes which occur during training could improve the monitoring process. This information could result in the prevention of stroke deterioration through improved feedback and improved management of the training load to maximise effective training time.

7.2.2. Sport Scientists

The identification of fifteen measureable variables (fourteen technique variables and swim time) using 2-D video analysis which show acute changes during high-intensity sets in Chapters 3 and 4 can result in the capacity to achieve more focused monitoring of technical factors during training. According to Coutts et al. (2014), sport scientists can become so engrossed and preoccupied with collecting data, that often they are not able to analyse it effectively nor provide the coach or athlete with any relevant feedback.

This thesis has provided a focused number of technique variables, measureable using a method applicable in training. This implies sport scientists could provide feedback which is of a high standard to athletes quickly using methods to which coaches can relate. This could help the daily practices of sport scientists, provide a better understanding of specific requirements, and guide coaches and training programmes more effectively. The identification of the barriers coaches perceive to the use of video analysis in Chapter 3 can assist sport scientists to work with coaches, and to develop the coach-sport scientist relationship.

7.2.3. Education Programs

Despite the range of literature pertaining to fatigue in sport (Ament and Verkerke, 2009), research regarding coaches' perceptions and management of this concept has not previously been conducted. Study three established that coaches' have a high consideration of fatigue during training, yet it is an area of inconsistency amongst coaches in terms of knowledge and methods. Identifying the gaps in coaches' current knowledge and the methods being used to monitor it could identify potential weak areas which require further development. In addition, information from on-going research, such as the individual and common changes in technique identified in Chapter 4, is also important in coach development and it is imperative that data of this nature are available to coaches. This has implications for the development of coach education programs to ensure they contain the relevant information required to ensure coaches' current knowledge on fatigue during training is up-to-date. It also proposes that this information should be disseminated during conferences and CPD courses to maintain coaching knowledge. Ensuring coaches have access to this information may aid the coaching and training performance, maximising the potential for athlete success.

7.2.4. Industry

Despite the barriers they perceive to accessing and using video analysis during training coaches place a high value on visual observation of fatigue during training due to its capacity to obtain and provide feedback to athletes quickly. In addition to this, coaches indicated a growing interest and willingness to use 2-D video analysis during training and by the end of the project a large number of coaches had either purchased, planned to purchase, or used a video camera as a monitoring tool during training. Thus, this research has implications for those involved in producing and selling 2-D video cameras and video analysis software. The continued development of 2-D video methods could result in more advanced pieces of technology becoming available to coaches and sport scientists for use in training situations. The outcomes of this research should be distributed to the video industry to see how video cameras and Dartfish software can be further improved, specifically for use by swimming coaches.

7.2.5. Researchers

One of the predominant findings of this thesis was the scarcity of research into the implementation of 2-D video analysis and the concept of fatigue specifically during training. Although this is a key part of athlete development, it is often overshadowed in research by other areas, particularly competitive performance as it is the penultimate outcome of sports performance. The training process is vital for optimal performance and highly important in developing the key skills necessary to compete. This research highlighted a wealth of areas which require further understanding and research. Potential avenues of research include: identifying other effects of fatigue during training, establishing the effect of video application into the coaches' daily routines, and establishing the long-term consequences of incorrect technical performance during high-intensity training.

7.2.6. Competitive Athletes

The enhanced knowledge, development of coach education programs, combined work of numerous individuals and enhanced feedback would be highly important to the athlete and their development. Study two revealed that acute technical changes could occur during a high-intensity swim session and validated the use of 2-D measures to identify these changes. The development of 2-D video analysis to monitor this (identified in Chapter 3) and enhance athlete feedback could have implications for swimmers training with poor technique, including a potential reduction in injury, muscular imbalances, and the development of bad technical habits which could result in a deteriorated race performance (Kluemper et al., 2006, Smith, 2003). This could have consequences for the overall training process and highlights the potential to aid athletic performance during competitions.

7.2.7. Implications of Findings Summary

By combining all of these factors and implications, this thesis shows the potential to reach a wide range of individuals involved in improving sports performance. Previous research has shown extensive differences between sport science and coaching in a number of aspects including: equipment use, knowledge, terminology, and perceptions of research (Kilic and Ince, 2015, Martindale and Nash, 2013, Williams and Kendall, 2007). Despite these challenges, the current work managed to begin bringing together coaches and sport scientists by assessing both disciplines in terms of the use of video technology and identifying:

- The measureable variables of technique, which are informative to coaches to monitor fatigue, using an applicable 2-D video analysis method available to both coaches and sport scientists.
- The perceived importance of fatigue during the training process by coaches.
- The barriers coaches have to using video technology during training.
- The successful implementation of video technology into a coaching situation (training session).
- Coaches' willingness to learn and request for further information and education regarding video analysis and fatigue.

Thus, the biggest implication of this research is its capacity to begin bringing sport scientists and coaches closer together with the focus of improving athletic performance. The degree to which this thesis has influenced these areas was briefly shown in the coaches' responses and comments. The coaches stated a keen interest in the topic of fatigue and the use of video cameras during training and requested additional education on this topic in the form of CPD courses. Further work is needed to identify the best way to provide this information and implement 2-D video methods into the coaching role.

7.3. CRITICAL REFLECTIONS AND FUTURE RESEARCH

The current research was necessary and essential to assess the differences between coaches and sport scientists and address this novel research problem. However, as a result of its exploratory nature, a number of limitations are apparent.

Firstly, as a result of investigating which variables could be assessed using 2-D video and analysis software a number of variables were excluded from analysis. This does not imply that the remaining variables do not change, nor that they cannot be observed by coaches. It does emphasise the need for further research into the biomechanical changes as a result of fatigue during training. Examples could include investigations into the remaining three technical strokes in swimming, or using 3-D video analysis methods to assess the technical changes. Although 3-D analysis is not as applicable and is difficult to use by coaches during training, its ability to measure a larger range of technical variables with a greater accuracy make its use in research highly relevant to the progression of knowledge in swimming biomechanics. Future research should involve analysing the effects of fatigue on swimming technique utilising 3-D video analysis methods to determine whether changes in other technical variables can be identified, and whether these variables can be measured using 2-D video methods or are observable by coaches.

Secondly, 2-D video analysis was applied in a coaching context; however, due to the exploratory nature, the present studies were not conducted specifically in a training environment. The second study took place under training-like conditions, but factors such as only one swimmer being tested at a time, and the set mimicking part of a training session and not lasting the same duration, may have influenced the results. In addition, the coaches' knowledge and methods they described in Study three, and the interviews conducted in Study four, took the coaches' practices and knowledge at their word. This was not formally assessed, or analysed, in an actual training environment or session. Future research could be conducted to investigate the technical changes which could occur during an actual high-intensity two hour swim training session, assess coaches' current knowledge of fatigue areas and what aspects coaches observe on poolside during an actual training session. This could include investigating whether observation differs from poolside and if alternative feedback methods using video can continue to aid observation and, therefore, coaches' feedback to athletes.

Thirdly, the present study focused on elite level swimmers and coaches working with national level athletes or higher due to the scarcity of previous research in this topic. There is a need to analyse at all levels of performance and coaching. Literature has shown that individual differences do occur between swimmers of different performance levels (Maglischo, 2003). As a result, differences in terms of technique may be apparent in less elite individuals, who are still training with large volumes which could be detrimental to performance (Alberty et al., 2009). Differences in observation have also been noted between coaches of different experience levels (Leas and Chi, 1993, Waters et al., 2014).

Therefore, future research could attempt to assess whether there are differences between performance level and changes in technique as a result of training, or in the technical variables coaches observe at different experience levels.

Fourthly, although this work focused on the biomechanical effects of fatigue during training and implementation of video to monitor this, it does not address the questions of 'why?': 'why does the technique change?' or 'why do coaches observe these technical factors?' Due to the vital role fatigue plays in the training process, such knowledge is essential to enhance the understanding of technical changes with fatigue, the observation and monitoring of fatigue during training and the application of sport science research to a coaching environment.

These topics for future research illuminate some of the areas which are yet to be adequately explored. However, the proposed future work would only be effective if applied in a sporting or coaching context. To understand the effects of such changes on technical performance and long-term consequences, liaising with other disciplines of sport science and sport performance is imperative to maximise the ability of the coach to provide effective feedback to athletes. Therefore I intend to follow up with institutions, including Scottish Swimming, British Swimming and the Institutes of Sport to distribute the results of this work to coaches and those individuals involved in sport performance. Results will be presented via publications, conference presentations and other accessible sources. I also intend to develop and offer CPD or workshops on a range of topics, including: fatigue and its role during training, the use of video technology (camera formats, data collection set up), the analysis of video footage, and the dissemination of data to athletes. The high level of interest and willingness of coaches to learn suggests that coaches are keen for such developments and further information.

While critically reflecting on this piece of research, I also realised an important aspect about myself as a researcher. On beginning this PhD journey, my focus was very much upon 3-D video analysis methods and stemmed from a scientific and sports biomechanics based background. Due to mitigating circumstances and equipment issues, the research question required adaptation and only used methods and equipment to which I had access, namely 2-D video analysis methods. This change identified a novel area in the realm of coaching and the use of video technology. It revealed a disparity between the sporting contexts I was familiar with, having been a coach and athlete myself, and the research focused realm of sports biomechanics I had studied. This was something I had not been aware of until this thesis. This work has resulted in me challenging my own views, as well as many aspects of sports performance and research methods I had previously assumed as ‘gospel’. On completing this work I have developed a new appreciation for the realm of applied sport science and a passion to continue promoting the application of research into a sporting context and bringing coaches and sport scientists closer together. The continued development and research into equipment used by coaches and by sport biomechanists, and the potential link and application of the two, will continue to aid and develop coaches’ and swimmers’ monitoring of fatigue during training. It is thought that more research using these methods could improve the acceptance of sport science by coaches and athletes. By continuing this approach, it could bring a balance by having both research and practice working to help and guide each other, rather than working individually, and this is greatly needed (Bishop, 2008a).

7.4. CONCLUSION

Video technology has developed to the point at which it can be applied for use during training by coaches, yet its application in swimming remains unknown, particularly in monitoring technical changes with fatigue during training. This is important to monitor due to the large role fatigue plays during high-intensity training in swimming and the effects it can have on technical performance.

In this thesis the implications of video technology to monitor fatigue during training were investigated. It identified that certain technical changes can occur during a highintensity training session and can be measured and monitored using 2-D video methods. Individual changes, both common and different, exist depending on the variable and technical style of the athlete; however, stroking parameters and swim velocity were found to change amongst all swimmers.

Coaches consider fatigue to play an important role during training and predominantly use visual observation or stopwatches to monitor fatigue due to perceived barriers of the application of video in a training environment. When using visual observation to observe technical changes with fatigue, coaches focused mainly on stroking parameters and leg glide time. Coaches’ perceptions of fatigue and video analysis appeared to play a key role in their monitoring of fatigue and the methods they choose. The small increase in the observation of technical changes with fatigue as a result of a brief video education intervention, combined with a keen interest in fatigue during training, suggest that video could be a helpful tool for coaches.

This innovative research has emphasised the importance of fatigue during the training process and a keen interest and willingness of coaches to learn and use video technology during training. This has implications for a wide range of individuals in terms of education, research and athlete performance. Further work is required to determine the best way to achieve this and to clarify the full potential of video and its application in monitoring fatigue during training for coaches of all working levels through continued research and work with coaches into relevant and practical research.

Chapter 7: Chapter Summary

What was already known about this topic?

- Swimming is a technique dependent sport.
- Fatigue has implications for the performance of technical actions in swimming.
- Sport scientists' and coaches' research, knowledge and methods differ in the monitoring process of athletes.

What new information does this chapter provide?

- Fourteen kinematic breaststroke technical variables and swim time can be measured using 2-D video analysis and have the potential for use to monitor fatigue during high-intensity training sets.
- Coaches' perceive the monitoring of fatigue as an important part of athlete development yet current education programs do not appear to cover the topic of fatigue sufficiently.
- 2-D video analysis can enhance coaches' abilities to consistently and accurately monitor changes in acute changes of breaststroke technique variables during training.
- Interventions using 2-D video analysis have the potential to improve coaches' capacity to observe acute technical changes with fatigue during training.
- 2-D video analysis may be the first tool to be used by both sport scientists and coaches, and has huge potential to begin bridging the gap that currently exists between these two groups as a tool to monitor fatigue. The research implies that there is a willingness to learn and use video analysis in coaching.

Bibliography

ABBISS, C. R. & LAURSEN, P. B. 2005. Models to explain fatigue during prolonged endurance cycling. *Sports Medicine,* 35, p. 865-898.

ACHTEN, J. & JEUKENDRUP, A. E. 2003. Heart rate monitoring. *Sports Medicine,* 33, p. 517-538.

ADRIAN, M. & COOPER, J. M. 1995. *The biomechanics of human movement,* University of Virginia, USA, Benchmark Press.

ALBERTY, M., POTDEVIN, F., DEKERLE, J., PELAYO, P., GORCE, P. & SIDNEY, M. 2008. Changes in swimming technique during time to exhaustion at freely chosen and controlled stroke rates. *Journal of Sports Sciences,* 26, p. 1191-1200.

ALBERTY, M., SIDNEY, M., HUOT-MARCHAND, F., HESPEL, J. & PELAYO, P. 2005. Intracyclic velocity variations and arm coordination during exhaustive exercise in front crawl stroke. *International Journal of Sports Medicine,* 26, p. 471-475.

ALBERTY, M., SIDNEY, M., PELAYO, P. & TOUSSAINT, H. 2009. Stroking characteristics during time to exhaustion tests. *Medicine and Science in Sports and Exercise,* 41, p. 637-644.

ALBERTY, M. R., POTDEVIN, F. P., DEKERLE, J., PELAYO, P. P. & SIDNEY, M. C. 2011.

Effect of stroke rate reduction on swimming technique during paced exercise.

The Journal of Strength & Conditioning Research, 25, p. 392-397.

ALLEY, L. E. 1952. An analysis of water resistance and propulsion in swimming the crawl stroke. *Research Quarterly. American Association for Health, Physical Education and Recreation,* 23, p. 253-270.

ALLISON, P. C. 1987. What and how preservice physical education teachers observe during an early field experience. *Research Quarterly for Exercise and Sport,* 58, p. 242-249.

AMENT, W. & VERKERKE, G. J. 2009. Exercise and fatigue. *Sports Medicine,* 39, p. 389422.

AMOROSO, A., SANDERSON, D. J. & HENNING, E. Kinematic and kinetic changes in cycling resulting from fatigue. Proceedings of the 17th meeting of the American society of biomechanics, 1993 Iowa City, USA. p. 157-158.

ANDERSON, R. Augmented feedback–the triptych conundrum. Proceeding of the XXVIII International Symposium of Biomechanics in Sports, 2010 Michigan, USA.

ARELLANO, R. & GAVILAN, P. Vortices and propulsion. SWIMMING: Applied proceedings of the XXII international symposium on biomechanics in sports, 1999 Perth, Australia. School of Biomedical and Sports Science, Edith Cowan University, p. 53-66.

AREND, S. & HIGGINS, J. R. 1976. A strategy for the classification, subjective analysis, and observation of human movement. *Journal of Human Movement Studies,* 2, p. 36-52.

ARROYO-TOLEDO, J., CLEMENTE, V. J., GONZALEZ-RAVE, J. M., RAMOS CAMPO, D. J. & SORTWELL, A. 2013. Comparison between traditional and reverse periodization: swimming performance and specific strength values.

International Journal of Swimming Kinetics, 2, p. 87-96.

ATKINSON, G. & NEVILL, A. M. 1998. Statistical methods for assessing measurement error (reliability) in variables relevant to sports medicine. *Sports Medicine,* 26, p. 217-238.

AUJOUANNET, Y. A., BONIFAZI, M., HINTZY, F., VUILLERME, N. & ROUARD, A. H. 2006. Effects of a high-intensity swim test on kinematic parameters in high-level athletes. *Applied Physiology, Nutrition, and Metabolism,* 31, p. 150-158.

BALILIONIS, G., NEPOCATYCH, S., ELLIS, C. M., RICHARDSON, M. T., NEGGERS, Y. H. & BISHOP, P. A. 2012. Effects of different types of warm-up on swimming performance, reaction time, and dive distance. *The Journal of Strength & Conditioning Research,* 26, p. 3297-3303.

BALSALOBRE-FERNÁNDEZ, C., TEJERO-GONZÁLEZ, C. M., DEL CAMPO-VECINO, J. & BAVARESCO, N. 2014. The concurrent validity and reliability of a low-cost, highspeed camera-based method for measuring the flight time of vertical jumps. *The Journal of Strength & Conditioning Research,* 28, p. 528-533.

BARBOSA, T. M., FERNANDES, R., KESKINEN, K., COLAÇO, P., CARDOSO, C., SILVA, J. & VILAS-BOAS, J. 2006. Evaluation of the energy expenditure in competitive swimming strokes. *International Journal of Sports Medicine,* 27, p. 894-899.

BARBOSA, T. M., FERNANDES, R., KESKINEN, K. & VILAS-BOAS, J. 2008. The influence of stroke mechanics into energy cost of elite swimmers. *European Journal of Applied Physiology,* 103, p. 139-149.

BARRETT, K. R. 1979. Observation for teaching and coaching. *Journal of Physical Education and Recreation,* 50, p. 23-25.

BARRETT, K. R. 1983. A hypothetical model of observing as a teaching skill. *Journal of Teaching in Physical Education,* 3, p. 22-31.

BARTLETT, R. 2002. *Sports biomechanics: reducing injury and improving performance,* New York, Routledge.

BARTLETT, R. 2007. *Introduction to sports biomechanics: Analysing human movement patterns,* New York, Routledge.

BARTLETT, R. M. 1997. Current issues in the mechanics of athletic activities. A position paper. *Journal of Biomechanics,* 30, p. 477-486.

BARTOLI, A., DALAL, N. & HORAUD, R. 2004. Motion panoramas. *Computer Animation and Virtual Worlds,* 15, p. 501-517.

BAUDRY, L., LEROY, D. & CHOLLET, D. 2006a. The effect of combined self-and expertmodelling on the performance of the double leg circle on the pommel horse.

Journal of Sports Sciences, 24, p. 1055-1063.

BAUDRY, L., LEROY, D., THOUVARECQ, R. & CHOLLET, D. 2006b. Auditory concurrent feedback benefits on the circle performed in gymnastics. *Journal of Sports Sciences,* 24, p. 149-156.

BEANLAND, E., MAIN, L. C., AISBETT, B., GASTIN, P. & NETTO, K. 2014. Validation of GPS and accelerometer technology in swimming. *Journal of Science and Medicine in Sport,* 17, p. 234-238.

BECKER, T. & HAVRILUK, R. 2006. Bilateral and anterior-posterior muscular imbalances in swimmers. *Biomechanics and Medicine in Swimming X. Portuguese Journal of Sport Sciences,* 6, p. 327-328.

BELL, L., M & INGLE, L. 2013. Psycho-physiological markers of overreaching and overtraining in endurance sports: a review of the evidence *Medicina Sportiva,* 17, p. 81-97.

BELOKOVSKY, V. & IVANCHENKO, E. A hydrokinetic apparatus for the study and improvement of leg movements in the breaststroke. Swimming II: proceedings of the Second International

Symposium on Biomechanics in Swimming, Brussels, Belgium, 1975. University Park Press, p. 64.

BERGER, M. A., DE GROOT, G. & HOLLANDER, A. P. 1995. Hydrodynamic drag and lift forces on human hand/arm models. *Journal of Biomechanics,* 28, p. 125-133.

BERTRAM, C., MARTENIUK, R. & GUADAGNOLI, M. 2007. On the use and misuse of video analysis. *International Journal of Sports Science and Coaching,* 2, p. 37-46.

BIELEC, G. & MAKAR, P. 2010. Variability in swimmers' individual kinematics parameters versus training loads. *Biology of Sport,* 27, p. 143.

BIRD, M. & HUDSON, J. 1990. Biomechanical observation: Visually accessible variables. *Biomechanics in sports VIII,* p. 219-223.

BISHOP, D. 2003. Warm up II. *Sports Medicine,* 33, p. 483-498.

BISHOP, D. 2008a. An applied research model for the sport sciences. *Sports Medicine,* 38, p. 253-263.

BISHOP, D. 2008b. www.pponline.co.uk [Online]. [Accessed 15.01.2015 2014].

BOBER, T. 1981. Biomechanical aspects of sports techniques. *Biomechanics VII,* p. 501509.

BOBER, T. & CZABANSKI, B. Changes in breaststroke techniques under different speed conditions. Second International Symposium on Biomechanics in Swimming, 1975 Brussels, Belgium. University Park Press, p. 188-193.

BONACCI, J., CHAPMAN, A., BLANCH, P. & VICENZINO, B. 2009. Neuromuscular adaptations to training, injury and passive interventions. *Sports Medicine,* 39, p. 903-921.

BOOROFF, M., NELSON, L. & POTRAC, P. 2015. A coach's political use of video-based feedback: a case study in elite-level academy soccer. *Journal of Sports Sciences,* p. 1-9.

BORRESEN, J. & LAMBERT, M. I. 2009. The quantification of training load, the training response and the effect on performance. *Sports Medicine,* 39, p. 779-795.

BREWIN, M. A. & KERWIN, D. G. 2003. Accuracy of scaling and DLT reconstruction techniques for planar motion analyses. *Journal of Applied Biomechanics,* 19, p. 79-88.

BUNN, J. W. 1972. *Scientific principles of coaching,* University of Michigan, Prentice Hall.

BURKE, L. M. 2008. Caffeine and sports performance. *Applied Physiology, Nutrition, and Metabolism,* 33, p. 1319-1334.

CAIRNS, S. P., KNICKER, A. J., THOMPSON, M. W. & SJØGAARD, G. 2005. Evaluation of models used to study neuromuscular fatigue. *Exercise and Sport Sciences Reviews,* 33, p. 9-16.

CALLAWAY, A., COBB, J. & JONES, I. 2009. A comparison of video and accelerometer based approaches applied to performance monitoring in swimming.

International Journal of Sports Science and Coaching, 4, p. 139-153.

CAPITÃO, F., LIMA, A., GONÇALVES, P., MOROUÇO, P., SILVA, M., FERNANDES, R. & VILAS-BOAS, J. 2006. Videogrametrically and velocimetrically assessed intracyclic variations of the velocity in breaststroke. *Biomechanics and Medicine in Swimming X,* p. 212-214.

CAPPAERT, J. 1996. Rotational balance about the center of mass in the breaststroke. *Swimming Science VII. London: E & FN SPON,* p. 29-33.

CAPPAERT, J. 1999. Biomechanics of swimming analysed by three-dimensional techniques. *Biomechanics and Medicine in Swimming,* 8, p. 141-145.

CARLING, C., WRIGHT, C., NELSON, L. J. & BRADLEY, P. S. 2014. Performance analysis in football: A critical review and implications for future research. *Journal of Sports Sciences,* 32, p. 2-7.

CARR, G. & CARR, G. A. 1997. *Mechanics of sport: a practitioner's guide,* Champaign, IL United States, Human kinetics.

CASA, D. J., CLARKSON, P. M. & ROBERTS, W. O. 2005. American College of Sports Medicine roundtable on hydration and physical activity: consensus statements.

Current Sports Medicine Reports, 4, p. 115-127.

CATY, V., AUJOUANNET, Y., HINTZY, F., BONIFAZI, M., CLARYS, J. & ROUARD, A. 2007.

Wrist stabilisation and forearm muscle coactivation during freestyle swimming. *Journal of Electromyography and Kinesiology,* 17, p. 285-291.

CHAN, C., YUNG, S., LIU, M. & CHAN, K. 2014. The change in lower limb kinematic and EMG activity of sprinting after soccer-specific fatigue among female soccer players. *British Journal of Sports Medicine,* 48, p. 578-578.

CHATARD, J.-C., GIROLD, S., CAUDAL, N., COSSOR, J. & MASON, B. 2003. Analysis of the 200 m events in the Sydney Olympic Games. *IX Biomechanics and Medicine in Swimming: University of Saint Etienne,* p. 261-264.

CHOLLET, D. 2007. Intra-cyclic distance per stroke phase, velocity fluctuations and acceleration time ratio of a breaststroker's hip: a comparison between elite and nonelite swimmers at different race paces. *International Journal of Sports Medicine,* 28, p. 140-147.

CHOLLET, D., CHALIES, S. & CHATARD, J. 2000. A new index of co-ordination for the crawl: description and usefulness. *International Journal of Sports Medicine,* 21, p. 54-59.

CHOLLET, D. & PELAYO, P. 1999. Effects of different methodologies in calculating stroke length in swimming. *Journal of Human Movement Studies,* 36, p. 127-136.

CHOLLET, D., PELAYO, P., DELAPLACE, C., TOURNY, C. & SIDNEY, M. 1997. Stroking characteristic variations in the 100-m freestyle for male swimmers of differing skill. *Perceptual and Motor Skills,* 85, p. 167-177.

CHOLLET, D., PELAYO, P., TOURNY, C. & SIDNEY, M. 1996. Comparative analysis of 100m and 200m events in the four strokes in top level swimmers. *Journal of Human Movement Studies,* 31, p. 25-38.

CHOLLET, D., SEIFERT, L., LEBLANC, H., BOULESTEIX, L. & CARTER, M. 2004. Evaluation of arm-leg coordination in flat breaststroke. *International Journal of Sports Medicine,* 25, p. 486-495.

CHRISTINA, K. A., WHITE, S. C. & GILCHRIST, L. A. 2001. Effect of localized muscle fatigue on vertical ground reaction forces and ankle joint motion during running. *Human Movement Science,* 20, p. 257-276.

CLARYS, J. 1979. Human morphology and hydrodynamics. *Swimming III,* 8, p. 3-41.

COE, R. It's the effect size, stupid: What effect size is and why it is important. British Educational Research Association annual conference, 2002 Exeter, UK. University of Durham, p. 1-18.

COLEMAN, S. G. 1999. Biomechanics and its application to coaching practice. *In:* CROSS, N. & LYLE, J. (eds.) *The coaching process: principles and practice for sport.* Oxford, UK: Butterworth-Heinemann.

COLMAN, V. & PERSYN, U. 1993. Trunk rotation, body waving and propulsion in breaststroke. *Journal of Human Movement Studies,* 24, p. 169-189.

COLMAN, V., PERSYN, U., DALY, D. & STIJNEN, V. 1998. A comparison of the intra-cyclic velocity variation in breaststroke swimmers with flat and undulating styles. *Journal of Sports Sciences,* 16, p. 653-665.

CONCEIÇÃO, A., SILVA, A. J., BARBOSA, T., KARSAI, I. & LOURO, H. 2014. Neuromuscular Fatigue during 200 M Breaststroke. *Journal of Sports Science & Medicine,* 13, p. 200-210.

COSTILL, D., MAGLISCHO, E. & RICHARDSON, A. 1992. *Swimming.,* London, UK, Blackwell Scientific Publications.

COUNSILMAN, J. E. 1948. A cinematographic analysis of the butterfly breaststroke. Unpublished: Master's thesis, University of Illinois at Urbana-Champaign.

COUTTS, A. J., ZAVORSKY, G. S., GALY, O., PYNE, D. B., GUY, J. H., EDWARDS, A. M., MURPHY, A. P., DUFFIELD, R., KELLETT, A. D. & REID, M. 2014. In the Age of Technology, Occam's Razor Still Applies. *International Journal of Sports Physiology and Performance,* 9, p. 2014-0353.

COVENTRY, E., BALL, K., PARRINGTON, L., AUGHEY, R. & MCKENNA, M. 2015. Kinematic effects of a short-term fatigue protocol on punt-kicking performance. *Journal of Sports Sciences*, p. 1-10.

CRAIG, A., BOOMER, W. & SKEHAN, P. 1988. Patterns of velocity in competitive breaststroke swimming. *Swimming Science V,* p. 73-77.

CRAIG, A. B. & PENDERGAST, D. R. 1979. Relationships of stroke rate, distance per stroke, and velocity in competitive swimming. *Medicine and Science in Sports and Exercise,* 11, p. 278-83.

CRAIG, A. B., SKEHAN, P. L., PAWELCZYK, J. A. & BOOMER, W. L. 1985. Velocity, stroke rate, and distance per stroke during elite swimming competition. *Medicine and Science in Sports and Exercise,* 17, p. 625-634.

CRAIG, A. B., TERMIN, B. & PENDERGAST, D. R. 2006. Simultaneous recordings of velocity and video during swimming. *Portuguese Journal of Sport Sciences,* 6, p. 32-35.

CUTTING, J. E. & KOZLOWSKI, L. T. 1977. Recognizing friends by their walk: Gait perception without familiarity cues. *Bulletin of The Psychonomic Society,* 9, p. 353-356.

CZABANSKI, B. & KOSZCYC, T. 1979. Relationship of stroke asymmetry and speed of breastroke swimming. *Swimming III,* p. 148-152.

D'ACQUISTO, L. & COSTILL, D. 1998. Relationship between intracyclic linear body velocity fluctuations, power, and sprint breaststroke performance. *Journal of Swimming Research,* 13, p. 8-14.

D'ACQUISTO, M. 1988. *Breaststroke economy, skill and performance study of breaststroke mechanics, using a computer based"*

Velocity. Video" system, Zentrum f. Wissenschaftsinformation, Körperkultur u. Sport.

DAVIS, J. M. & BAILEY, S. P. 1997. Possible mechanisms of central nervous system fatigue during exercise. *Medicine and Science in Sports and Exercise,* 29, p. 45-57.

DEBANNE, T. & FONTAYNE, P. 2009. A study of a successful experienced elite handball coach's cognitive processes in competition situations. *International Journal of Sports Science and Coaching,* 4, p. 1-16.

DEKERLE, J., NESI, X., LEFEVRE, T., DEPRETZ, S., SIDNEY, M., MARCHAND, F. H. & PELAYO, P. 2004. Stroking parameters in front crawl swimming and maximal lactate steady state speed. *International Journal of Sports Medicine,* 26, p. 53-58.

DELEXTRAT, A., TRICOT, V., BERNARD, T., VERCRUYSSEN, F., HAUSSWIRTH, C. & BRISSWALTER, J. 2005. Modification of cycling biomechanics during a swim-tocycle trial. *Journal of Applied Biomechanics,* 21, p. 297-308.

DERRICK, T. R., DEREU, D. & MCLEAN, S. P. 2002. Impacts and kinematic adjustments during an exhaustive run. *Medicine and Science in Sports and Exercise,* 34, p. 998-1002.

DESCHODT, V., ARSAC, L. & ROUARD, A. 1999. Relative contribution of arms and legs in humans to propulsion in 25-m sprint front-crawl swimming. *European Journal of Applied Physiology and Occupational Physiology,* 80, p. 192-199.

DESCHODT, V., ROUARD, A. & MONTEIL, K. 1996. Relationships between the three coordinates of the upper limb joints with swimming velocity. *Biomechanics and medicine in swimming VII. London: E & FN Spon,* p. 52-58.

DINGENEN, B., MALFAIT, B., VANRENTERGHEM, J., VERSCHUEREN, S. M. & STAES, F. F. 2014. The reliability and validity of the measurement of lateral trunk motion in two-dimensional video analysis during unipodal functional screening tests in elite female athletes. *Physical Therapy in Sport,* 15, p. 117-123.

DIXON, M., WONG, R., CSICSERY, K., POPICH, A., KLASSEN, D., MEHNDIRATTA, V. & HIGGINSON, B. The effects of aerobic fatigue on joint kineatics and torque production in cyclists. International Journal of Exercise Science: Conference Proceedings, 2013. p. 6.

DORGO, S. 2009. Unfolding the practical knowledge of an expert strength and conditioning coach. *International Journal of Sports Science and Coaching,* 4, p. 17-30.

DOUWES, M. & DUL, J. Validity and reliability of estimating body angles by direct and indirect observations. Designing for everyone. Proceedings of the 11th Congress of the International Ergonomics Association, 1991. p. 885-7.

EAST, D. 1970. Swimming: an analysis of stroke frequency, stroke length and performance. *New Zealand Journal of Health, Physical Education and Recreation,* 3, p. 16-27.

ELBE, A.-M., RASMUSSEN, C. P., NIELSEN, G. & NORDSBORG, N. B. 2015. High intensity and reduced volume training attenuates stress and recovery levels in elite swimmers. *European Journal of Sport Science*, p. 1-6.

ELLIOTT, B. 1999. Biomechanics: an integral part of sport science and sport medicine. *Journal of Science and Medicine in Sport,* 2, p. 299-310.

ELLIOTT, B. & BARTLETT, R. 2006. Sports Biomechanics: Does it have a Role in Coaching? *International Journal of Sports Science and Coaching,* 1, 177-183.

ELTOUKHY, M., ASFOUR, S., THOMPSON, C. & LATTA, L. 2012. Evaluation of the performance of digital video analysis of human motion: Dartfish tracking system. *International Journal of Scientific and Engineering Research (IJSER),* 3, p. 1-6.

ENOKA, R. M. & DUCHATEAU, J. 2008. Muscle fatigue: what, why and how it influences muscle function. *The Journal of Physiology,* 586, p. 11-23.

ENOKA, R. M. & STUART, D. G. 1992. Neurobiology of muscle fatigue. *Journal of Applied Physiology,* 72, p. 1631-1648.

ERICKSON, K., BRUNER, M., MACDONALD, D. & CÔTÉ, J. 2008. Gaining insight into actual and preferred sources of coaching knowledge. *International journal of sports science and coaching,* 3, 527-538.

ERICSSON, K. A., KRAMPE, R. T. & TESCH-RÖMER, C. 1993. The role of deliberate practice in the acquisition of expert performance. *Psychological Review,* 100, p. 363-406.

FEDEROLF, P., REID, R., GILGIEN, M., HAUGEN, P. & SMITH, G. 2014. The application of principal component analysis to quantify technique in sports. *Scandinavian Journal of Medicine & Science in Sports,* 24, p. 491-499.

FIELD, A. 2009. *Discovering statistics using SPSS,* London, UK, Sage Publications Ltd. FIGUEIREDO, P., SANDERS, R., GORSKI, T., VILAS-BOAS, J. & FERNANDES, R. 2013. Kinematic and electromyographic changes during 200 m front crawl at race pace. *International Journal of Sports Medicine,* 34, p. 49-55.

FITTS, R. H. 2008. The cross-bridge cycle and skeletal muscle fatigue. *Journal of Applied Physiology,* 104, p. 551-558.

FLEMING, P., YOUNG, C., DIXON, S. & CARRÉ, M. 2010. Athlete and coach perceptions of technology needs for evaluating running performance. *Sports Engineering,* 13, p. 1-18.

FORESTIER, N. & NOUGIER, V. 1998. The effects of muscular fatigue on the coordination of a multijoint movement in human. *Neuroscience Letters,* 252, p. 187-190.

FRANKS, I. M. 1993. The effects of experience on the detection and location of performance differences in a gymnastic technique. *Research Quarterly for Exercise and Sport,* 64, p. 227-231.

FRANKS, I. M. & MILLER, G. 1991. Training coaches to observe and remember. *Journal of Sports Sciences,* 9, p. 285-297.

FULTON, S. K., PYNE, D. B. & BURKETT, B. 2009. Validity and reliability of kick count and rate in freestyle using inertial sensor technology. *Journal of Sports Sciences,* 27, p. 1051-1058.

GABBETT, T. J. 2008. Influence of fatigue on tackling technique in rugby league players. *The Journal of Strength & Conditioning Research,* 22, p. 625-632.

GABRIEL, D. A., BASFORD, J. R. & AN, K.-N. 2001a. Neural adaptations to fatigue: implications for muscle strength and training. *Medicine and Science in Sports and Exercise,* 33, p. 1354-1360.

GABRIEL, D. A., BASFORD, J. R. & AN, K.-N. 2001b. Neural adaptations to fatigue: implications for muscle strength and training. *Medicine and science in sports and exercise,* 33, 1354-1360.

GARHAMMER, J. & NEWTON, H. 2013. Applied Video Analysis For Coaches: Weightlifting Examples. *International Journal of Sports Science and Coaching,* 8, p. 581-594.

GERLACH, K. E., WHITE, S. C., BURTON, H. W., DORN, J. M., LEDDY, J. J. & HORVATH, P. J. 2005. Kinetic changes with fatigue and relationship to injury in female runners. *Medicine and Science in Sports and Exercise,* 37, p. 657-663.

GIBSON, A. S. C. & NOAKES, T. 2004. Evidence for complex system integration and dynamic neural regulation of skeletal muscle recruitment during exercise in humans. *British Journal of Sports Medicine,* 38, p. 797-806.

GLAISTER, M. 2005. Multiple sprint work. *Sports Medicine,* 35, p. 757-777.

GOULD, D., GUINAN, D., GREENLEAF, C. & CHUNG, Y. 2002. A survey of US Olympic coaches: Variables perceived to have influenced athlete performances and coach effectiveness. *The Sport Psychologist,* 16, p. 229-250.

GRACE, T. 1985. Muscle imbalance and extremity injury. A perplexing relationship. *Sports Medicine (Auckland, NZ),* 2, p. 77.

GRANT, M. A., MCCULLICK, B. A., SCHEMPP, P. G. & GRANT, J. T. 2012. Experts' Content Knowledge of Fundamentals. *International Journal of Sports Science and Coaching,* 7, p. 399-410.

GRATTON, C. & JONES, I. 2010. *Research methods for sports studies,* New York, USA, Taylor & Francis.

GREEN, H. 1997. Mechanisms of muscle fatigue in intense exercise. *Journal of Sports Sciences,* 15, p. 247-256.

GUADAGNOLI, M., HOLCOMB, W. & DAVIS, M. 2002. The efficacy of video feedback for learning the golf swing. *Journal of Sports Sciences,* 20, p. 615-622.

HALSON, S. L., BRIDGE, M. W., MEEUSEN, R., BUSSCHAERT, B., GLEESON, M., JONES, D. A. & JEUKENDRUP, A. E. 2002. Time course of performance changes and fatigue markers during intensified training in trained cyclists. *Journal of Applied Physiology,* 93, p. 947-956.

HALSON, S. L. & JEUKENDRUP, A. E. 2004. Does overtraining exist? *Sports Medicine,* 34, p. 967-981.

HANDFORD, C., DAVIDS, K., BENNETT, S. & BUTTON, C. 1997. Skill acquisition in sport: Some applications of an evolving practice ecology. *Journal of Sports Sciences,* 15, p. 621-640.

HANNULA, D. 2003. *Coaching swimming successfully,* USA, Human Kinetics.

HARS, M. & CALMELS, C. 2007. Observation of elite gymnastic performance: Processes and perceived functions of observation. *Psychology of Sport and Exercise,* 8, p. 337-354.

HAUGEN, T. A., TØNNESSEN, E. & SEILER, S. K. 2012. The difference is in the start: impact of timing and start procedure on sprint running performance. *The Journal of Strength & Conditioning Research,* 26, p. 473-479.

HAY, J. G. 1993. *The biomechanics of sports techniques,* United States, Prentice Hall Inc.

HAY, J. G. & REID, G. 1982. *Anatomy, Mechanics, and Human Motion,,* Englewood Cliffs, NJ, Prentice-Hall.

HEBERT, E. P. & LANDIN, D. 1994. Effects of a learning model and augmented feedback on tennis skill acquisition. *Research Quarterly for Exercise and Sport,* 65, p. 250257.

HELLARD, P., DEKERLE, J., AVALOS, M., CAUDAL, N., KNOPP, M. & HAUSSWIRTH, C. 2008. Kinematic measures and stroke rate variability in elite female 200-m swimmers in the four swimming techniques: Athens 2004 Olympic semifinalists and French National 2004 Championship semi-finalists. *Journal of Sports Sciences,* 26, p. 35-46.

HERNÁNDEZ, F. J. M., ROMERO, F. Á., VAÍLLO, R. R. & DEL CAMPO, V. L. 2006. Visual behaviour of tennis coaches in a court and video-based conditions. *International Journal of Sport Science,* 5, p. 29-41.

HINCKSON, E. A. & HOPKINS, W. G. 2005. Reliability of time to exhaustion analyzed with critical-power and log-log modeling. *Medicine and Science in Sports and Exercise,* 37, p. 696-701.

HOCHMUTH, G. 1984. *Biomechanics of athletic movement,* Sportverlag.

HODGES, N. J., CHUA, R. & FRANKS, I. M. 2003. The role of video in facilitating perception and action of a novel coordination movement. *Journal of Motor Behavior,* 35, p. 247-260.

HODGES, N. J. & FRANKS, I. M. 2002. Modelling coaching practice: the role of instruction and demonstration. *Journal of Sports Sciences,* 20, p. 793-811.

HODGES, N. J., WILLIAMS, A. M., HAYES, S. J. & BRESLIN, G. 2007. What is modelled during observational learning? *Journal of Sports Sciences,* 25, p. 531-545. HOLLMAN, J. H., GINOS, B. E., KOZUCHOWSKI, J., VAUGHN, A. S., KRAUSE, D. A. & YOUDAS, J. W. 2009. Relationships between knee valgus, hip-muscle strength, and hip-muscle recruitment during a single-limb step-down. *Journal of Sport Rehabilitation,* 18, p. 104-117.

HOPKINS, W. G. 2000. Measures of reliability in sports medicine and science. *Sports Medicine,* 30, p. 1-15.

HOPKINS, W. G. 2002. *A new View of statistics* [Online]. [Accessed 15.01.14 2014].

HOPKINS, W. G. 2004. How to interpret changes in an athletic performance test. *Sportscience,* 8, p. 1-7.

HORNERY, D. J., FARROW, D., MUJIKA, I. & YOUNG, W. 2007. Fatigue in tennis. *Sports Medicine,* 37, p. 199-212.

HORTON, P. 2015. The role of the coach. *Practical Sports Coaching,* p. 3-15.

HOWELL, D. 2012. *Statistical methods for psychology,* Cengage Learning.

HUDSON, J. L. The value visual variables in biomechanical analysis. Proceedings of the VI international symposium of the international society of biomechanics in sports, 1990 Prague, Czechoslovakia. p. 499-509.

HUDSON, J. L. 1995. Core concepts of kinesiology. *Journal of Physical Education, Recreation & Dance,* 66, p. 54-60.

HUGHES, M. & FRANKS, I. 2007. *The essentials of performance analysis: an introduction,* New York, United States, Routledge.

HUGHES, M. D. & BARTLETT, R. M. 2002. The use of performance indicators in performance analysis. *Journal of Sports Sciences,* 20, p. 739-754.

HUNTER, S. K., DUCHATEAU, J. & ENOKA, R. M. 2004. Muscle fatigue and the mechanisms of task failure. *Exercise and Sport Sciences Reviews,* 32, p. 44-49.

IMWOLD, C. H. & HOFFMAN, S. J. 1983. Visual recognition of a gymnastics skill by experienced and inexperienced instructors. *Research Quarterly for Exercise and Sport,* 54, p. 149-155.

IRWIN, G. & KERWIN, D. G. 2005. Gymnastics: Biomechanical similarities of progressions for the longswing on high bar. *Sports Biomechanics,* 4, 163-178.

JOHANSSON, G. 1975. Visual motion perception. *Scientific American,* 232, p. 76-88.

JONES, R. L. & TURNER, P. 2006. Teaching coaches to coach holistically: Can ProblemBased Learning (PBL) help? *Physical Education and Sport Pedagogy,* 11, p. 181202.

JUDGE, L., HUNTER, I. & GILREATH, E. 2008. Using sport science to improve coaching: A case study of the American record holder in the women's hammer throw. *International Journal of Sports Science and Coaching,* 3, 477-488.

JURIMAE, J., HALJASTE, K., CICCHELLA, A., LATT, E., PURGE, P., LEPPIK, A. & JURIMAE, T. 2007. Analysis of swimming performance from physical, physiological, and biomechanical parameters in young swimmers. *Pediatric Exercise Science,* 19, p. 70-81.

JUSTHAM, L., SLAWSON, S., WEST, A., CONWAY, P., CAINE, M. & HARRISON, R. 2008. Enabling technologies for robust performance monitoring (P10). *The Engineering of Sport 7.* Paris, France: Springer.

KAY, D., MARINO, F. E., CANNON, J., GIBSON, A. S. C., LAMBERT, M. I. & NOAKES, T. D. 2001. Evidence for neuromuscular fatigue during high-intensity cycling in warm, humid conditions. *European Journal of Applied Physiology,* 84, 115-121.

KELLIS, E., KATIS, A. & VRABAS, I. 2006. Effects of an intermittent exercise fatigue protocol on biomechanics of soccer kick performance. *Scandinavian Journal of Medicine & Science in Sports,* 16, p. 334-344.

KELLY, J. 2007. *Time-course of kinetic and kinematic changes in cyclists' pedalling technique with fatigue.* PhD, University of Edinburgh.

KENNEDY, P., BROWN, P., CHENGALUR, S. N. & NELSON, R. C. 1990. Analysis of Male and Female Olympic Swimmers in the 100-Meter Events. *International Journal of Sport Biomechanics,* 6, p. 187-197.

KENT, M. & ATHA, J. Selected critical transient body positions in breast stroke and their influence upon water resistance. 1st International symposium of biomechanics in swimming. Eds. L. Lewillie and JP Clarys. Bruxelles, 1971. p. 125.

KERWIN, D. G. & IRWIN, G. 2008. Biomechanics for coaches. *In:* JONES, R. L., HUGHES, M. & K, K. (eds.) *An introduction to sports coaching: From Science and theory to practice.* New York, United States: Routledge.

KESKINEN, K. L. & KOMI, P. V. 1993. Stroking Characteristics of Front Crawl Swimming During Exercise. *Journal of Applied Biomechanics,* 9, p. 219-226.

KILIC, K. & INCE, M. L. 2015. Use of Sports Science Knowledge by Turkish Coaches. *International Journal of Exercise Science,* 8, p. 21-37.

KIRK, D., MACDONALD, D. & O'SULLIVAN, M. 2006. *Handbook of physical education,* London, UK, Sage Publications Ltd.

KIRMIZIBAYRAK, C., HONORIO, J., JIANG, X., MARK, R. & HAHN, J. K. 2011. Digital analysis and visualization of swimming motion. *International Journal of Virtual Reality,* 10, p. 9-16.

KLINE, C. E., DURSTINE, J. L., DAVIS, J. M., MOORE, T. A., DEVLIN, T. M., ZIELINSKI, M. R. & YOUNGSTEDT, S. D. 2007. Circadian variation in swim performance. *Journal of Applied physiology,* 102, p. 641-649.

KLUEMPER, M., UHL, T. & HAZELRIGG, H. 2006. Effect of stretching and strengthening shoulder muscles on forward shoulder posture in competitive swimmers. *Journal of Sport Rehabilitation,* 15, p. 58-70.

KNICKER, A. J., RENSHAW, I., OLDHAM, A. R. & CAIRNS, S. P. 2011. Interactive processes link the multiple symptoms of fatigue in sport competition. *Sports Medicine,* 41, p. 307-328.

KNUDSON, D. 1999. Validity and reliability of visual ratings of the vertical jump. *Perceptual and Motor Skills,* 89, p. 642-648.

KNUDSON, D. 2007. Qualitative biomechanical principles for application in coaching. *Sports Biomechanics,* 6, p. 109-118.

KNUDSON, D., ELLIOTT, B. & HAMILL, J. 2014. Proposing application of results in sport and exercise research reports. *Sports Biomechanics,* 13, p. 195-203.

KNUDSON, D. V. & MORRISON, C. S. 2002. *Qualitative Analysis of Human Movement,* United States, Human Kinetics.

KOLMOGOROV, S. V., RUMYANTSEVA, O., GORDON, B. & CAPPAERT, J. 1997. Hydrodynamic characteristics of competitive swimmers of different genders and performance levels. *Journal of Applied Biomechanics,* 13, p. 88-97.

KRAEMER, E. F. 2009. *Hip and Knee Positions During the Curling Delivery.* UNIVERSITY OF WISCONSIN-LA CROSSE.

KREIGHBAUM, E. & BARTHELS, K. 1996. *Biomechanics: A qualitative approach for studying human movement,* London, UK, Allyn and Bacon.

KROSSHAUG, T., NAKAMAE, A., BODEN, B., ENGEBRETSEN, L., SMITH, G., SLAUTERBECK, J., HEWETT, T. E. & BAHR, R. 2007. Estimating 3D joint kinematics from video sequences of running and cutting maneuvers—assessing the accuracy of simple visual inspection. *Gait & Posture,* 26, p. 378-385.

KRUCHOSKY, E. P. 1954. *A Performance Analysis of Drag and Propulsion in Swimming Three Selected Forms of the Back Crawl Stroke.* State University of Iowa.

LAFFITE, L. P., VILAS-BOAS, J. P., DEMARLE, A., SILVA, J., FERNANDES, R. & LOUISE BILLAT, V. 2004. Changes in physiological and stroke parameters during a maximal 400-m free swimming test in elite swimmers. *Canadian Journal of Applied Physiology,* 29, p. S17-S31.

LAMBERT, E., GIBSON, A. S. C. & NOAKES, T. 2005. Complex systems model of fatigue: integrative homoeostatic control of peripheral physiological systems during exercise in humans. *British Journal of Sports Medicine,* 39, p. 52-62.

LATASH, M. L. 2008. *Neurophysiological basis of movement,* United States, Human Kinetics.

LEAS, R. R. & CHI, M. T. 1993. Analyzing diagnostic expertise of competitive swimming coaches. *Advances in Psychology,* 102, p. 75-94.

LEBLANC, H., SEIFERT, L. & CHOLLET, D. 2009. Arm-leg coordination in recreational and competitive breaststroke swimmers. *Journal of Science and Medicine in Sport,* 12, p. 352-356.

LEBLANC, H., SEIFERT, L. & CHOLLET, D. 2010. Does floatation influence breaststroke technique. *Journal of Applied Biomechanics,* 26, p. 150-158.

LEBLANC, H., SEIFERT, L., TOURNY-CHOLLET, C. & CHOLLET, D. 2007. Velocity variations in breaststroke swimmers of different competitive levels. *International Journal of Sports Medicine,* 28, p. 140-147.

LEES, A. 1999. Biomechanical assessment of individual sports for improved performance. *Sports Medicine,* 28, p. 299-305.

LEES, A. 2002. Technique analysis in sports: a critical review. *Journal of Sports Sciences,* 20, p. 813-828.

LIEBERMANN, D. G., KATZ, L., HUGHES, M. D., BARTLETT, R. M., MCCLEMENTS, J. & FRANKS, I. M. 2002. Advances in the application of information technology to sport performance. *Journal of Sports Sciences,* 20, p. 755-769.

LOETZ, C., REISCHLE, K. & SCHMITT, G. 1988. The evaluation of highly skilled swimmers via quantitative and qualitative analysis. *Swimming Science V. Human Kinetics Books,* p. 361-367.

LYLE, J. 1999. The coaching process: an overview. *In:* CROSS, N. & LYLE, J. (eds.) *The coaching process: principles and practice for sport.* Oxford, UK: ButterworthHeinemann.

MACKENZIE, R. & CUSHION, C. 2013. Performance analysis in football: A critical review and implications for future research. *Journal of Sports Sciences,* 31, p. 639-676.

MAGILL, R. A. 2004. *Motor learning and control: Concepts and applications,* McGraw-Hill Education.

MAGLISCHO, E. W. 1982. *Swimming faster: a comprehensive guide to the science of swimming,* WCB/McGraw-Hill.

MAGLISCHO, E. W. 2003. *Swimming fastest,* United States, Human Kinetics.

MANLEY, P. & ATHA, J. 1992. Intra-stroke velocity fluctuations in paced breaststroke swimming. *Biomechanics and Medicine in Swimming VI. London: E & FN Spon,* p. 151-160.

MARCORA, S. M., STAIANO, W. & MANNING, V. 2009. Mental fatigue impairs physical performance in humans. *Journal of Applied Physiology,* 106, p. 857-864.

MARINHO, D., BOAS, J. P. V., KESKINEN, K., RODRIGUEZ, F., SOARES, S., CARMO, C., VILAR, S. & FERNANDES, R. 2006. Behaviour of the kinematic parameters during a time to exhaustion test at VO2max in elite swimmers. *Journal of Human Movement Studies,* 51, p. 1-10.

MARINHO, D. A., ROUBOA, A. I., ALVES, F. B., VILAS-BOAS, J. P., MACHADO, L., REIS, V. M. & SILVA, A. J. 2009. Hydrodynamic analysis of different thumb positions in swimming. *Journal of Sports Science & Medicine,* 8, p. 58-66.

MARINO, F. E., GARD, M. & DRINKWATER, E. J. 2009. The limits to exercise performance and the future of fatigue research. *British journal of sports medicine*, bjsports67611.

MARTINDALE, R. & NASH, C. 2013. Sport science relevance and application: Perceptions of UK coaches. *Journal of Sports Sciences,* 31, p. 807-819.

MARTINS-SILVA, A., ALVES, F. & GOMES-PEREIRA, J. Fatigue related technical changes in 3-D directional components of hand velocity in butterfly swimming during a 200 m maximum trial: a pilot study. Proceedings of the XII FINA World Congress on Sports Medicine, 1997 Goteburg, Sweden. p. 463-467.

MASON, B. & COSSOR, J. What can we learn from competition analysis at the 1999 Pan Pacific Swimming Championships. Proceedings of XVIII International Symposium on Biomechanics in Sports–Applied Program, 2000 Hong Kong. 7582.

MATOS, N. F., WINSLEY, R. J. & WILLIAMS, C. A. 2011. Prevalence of nonfunctional overreaching/overtraining in young English athletes. *Medicine and Science in Sports and Exercise,* 43, p. 1287-94.

MAUGHAN, R. 2012. Thermoregulatory aspects of performance. *Experimental Physiology,* 97, p. 325-326.

MCGARRY, T., O'DONOGHUE, P., DE EIRA SAMPAIO, A. J. & SAMPAIO, J. 2013. *Routledge handbook of sports performance analysis,* New York, United States, Routledge.

MCKENNA, M. J., BANGSBO, J. & RENAUD, J.-M. 2008. Muscle K+, Na+, and Cl− disturbances and Na+-K+ pump inactivation: implications for fatigue. *Journal of Applied Physiology,* 104, p. 288-295.

MCMORRIS, T. 2014. *Acquisition and performance of sports skills,* UK, John Wiley & Sons.

MCPHERSON, M. 1990. A systematic approach to skill analysis. *Sports Science Periodical on Research and Technology in Sport,* 11, p. 1-10.

MEEUSEN, R., DUCLOS, M., FOSTER, C., FRY, A., GLEESON, M., NIEMAN, D., RAGLIN, J., RIETJENS, G., STEINACKER, J. & URHAUSEN, A. 2013. Prevention, diagnosis, and treatment of the overtraining syndrome: joint consensus statement of the European College of Sport Science and the American College of Sports Medicine. *Medicine and Science in Sports and Exercise,* 45, p. 186-205.

METS, T., VIITASALO, J., MONONEN, K. & KONTTINEN, N. 2003. Reproducibility of coaches' observations in running target shooting performance. *Kinesiology,* 35, p. 201-209.

MIER, C. M. 2011. Accuracy and feasibility of video analysis for assessing hamstring flexibility and validity of the sit-and-reach test. *Research Quarterly for Exercise and Sport,* 82, p. 617-623.

MILLER, A. & CALLISTER, R. 2009. Reliable lower limb musculoskeletal profiling using easily operated, portable equipment. *Physical Therapy in Sport,* 10, p. 30-37.

MINETTI, A. E., MACHTSIRAS, G. & MASTERS, J. C. 2009. The optimum finger spacing in human swimming. *Journal of Biomechanics,* 42, p. 2188-2190.

MIYASHITA, M. An analysis of fluctuations of swimming speed. Proceedings of the First International Symposium on Biomechanics in Swimming waterpolo and Diving, 1971. p. 53-57.

MONTEIL, K. 1994. Free swimming versus paddles swimming in front crawl. *Journal of Human Movement Studies,* 27, p. 89.

MONTEIL, K., ROUARD, A., DUFOUR, A., CAPPAERT, J. & TROUP, J. 1996. Front crawl stroke phases: discriminating kinematic and kinetic parameters. *Biomechanics and medicine in swimming VII. London: E & FN Spon,* p. 45-51.

MOORE, W. E. & STEVENSON, J. R. 2010. Understanding trust in the performance of complex automatic sport skills. *The Sport Psychologist,* 5, p. 281-289.

MORENO, F. J., SAAVEDRA, J. M., SABIDO, R., LUIS, V. & REINA, R. 2006. Visual search strategies of experience and nonexperienced swimming coaches. *Perceptual and Motor Skills,* 103, p. 861-872.

MORIN, J.-B., SAMOZINO, P. & MILLET, G. Y. 2011. Changes in running kinematics, kinetics, and spring-mass behavior over a 24-h run. *Medicine and Science in Sports and Exercise,* 43, p. 829-836.

MORRISON, C. S., CLAYBURN, C., KNUDSON, D. & HAYWOOD, P. 2005. Accuracy of visual estimates of joint angle and angular velocity using criterion movements. *Perceptual and Motor Skills,* 100, p. 599-606.

MUNRO, A., HERRINGTON, L. & CAROLAN, M. 2012. Reliability of two-dimensional video assessment of frontal plane dynamic knee valgus during common athletic screening tasks. *Journal of Sport Rehabilitation,* 21, p. 7-11.

MYTTON, G. J., ARCHER, D. T., THOMPSON, K. G., RENFREE, A. & ST CLAIR GIBSON, A. 2013. Validity and Reliability of a 1500-m Lap-Time Collection Method Using Public Videos. *International Journal of Sports Physiology and Performance,* 8, p. 692-694.

NASH, C. & SPROULE, J. 2011. Insights into experiences: Reflections of an expert and novice coach. *International Journal of Sports Science and Coaching,* 6, p. 149162.

NASH, C. & SPROULE, J. 2012. Coaches perceptions of their coach education experiences. *International Journal of Sport Psychology,* 43, p. 33.

NASH, C. S. 2008. *The role of coach education in the development of expertise in coaching.* The University of Edinburgh.

NASH, C. S. & SPROULE, J. 2009. Career development of expert coaches. *International Journal of Sports Science and Coaching,* 4, p. 121-138.

NASH, C. S., SPROULE, J. & HORTON, P. 2011. Excellence in coaching: The art and skill of elite practitioners. *Research Quarterly for Exercise and Sport,* 82, p. 229-238.

NELSON, L. J. & GROOM, R. 2012. The analysis of athletic performance: some practical and philosophical considerations. *Sport, Education and Society,* 17, p. 687-701.

NEMESSURI, M. & VADAY, M. 1971. Breaststroke motor pattern. *Swimming Science I,* p. 161-165.

NEWELL, K. M. 1976. Knowledge of results and motor learning. *Exercise and Sport Sciences Reviews,* 4, p. 195-228.

NEZHAD, V. S., RAHIMI, A. & SARSHIN, A. 2015. Effect of fatigue on body joints kinematic in 3 point jump shot among basketball player young boys. *Biological Forum – An International Journal,* 7, p. 1910-1914.

NIELSEN, O. B., PAOLI, F. & OVERGAARD, K. 2001. Protective effects of lactic acid on force production in rat skeletal muscle. *The Journal of Physiology,* 536, p. 161166.

NIKODELIS, T., KOLLIAS, I. & HATZITAKI, V. 2005. Bilateral inter-arm coordination in freestyle swimming: Effect of skill level and swimming speed. *Journal of Sports Sciences,* 23, p. 737-745.

NIMZ, R., RADER, U., WILKE, K. & SKIPKA, W. 1988. The relationship of anthropometric measures to different types of breaststroke kicks. *Swimming Science V,* p. 115119.

NOAKES, T. 2000. Physiological models to understand exercise fatigue and the adaptations that predict or enhance athletic performance. *Scandinavian Journal of Medicine and Science in Sports,* 10, p. 123-145.

NOAKES, T. & GIBSON, A. S. C. 2004. Logical limitations to the "catastrophe" models of fatigue during exercise in humans. *British Journal of Sports Medicine,* 38, p. 648649.

NOAKES, T. D., PELTONEN, J. E. & RUSKO, H. K. 2001. Evidence that a central governor regulates exercise performance during acute hypoxia and hyperoxia. *Journal of Experimental Biology,* 204, p. 3225-3234.

NORRIS, B. S. & OLSON, S. L. 2011. Concurrent validity and reliability of twodimensional video analysis of hip and knee joint motion during mechanical lifting. *Physiotherapy Theory and Practice,* 27, p. 521-530.

NOSAKA, K., LAVENDER, A., NEWTON, M. & SACCO, P. 2003. Muscle damage in resistance training. *International Journal of Sport and Health Science,* 1, p. 1-8.

NYBO, L. 2008. Hyperthermia and fatigue. *Journal of Applied Physiology,* 104, p. 871878.

OLIVEIRA, N., SAUNDERS, D. H. & SANDERS, R. H. 2015. Fatigue Induced Changes in Eggbeater Kick Kinematics Affect Performance and Risk of Injury. *International Journal of Sports Physiology and Performance,* Epud ahead of print.

OZDOĞAN, Y. & OZCELIK, A. O. 2011. Evaluation of the nutrition knowledge of sports department students of universities. *Journal of the International Society of Sports Nutrition,* 5, p. 11.

PAI, Y. C., HAY, J. G. & WILSON, B. D. 1984. Stroking techniques of elite swimmers. *Journal of Sports Sciences,* 2, p. 225-239.

PARSONS, J. L. & ALEXANDER, M. J. 2012. Modifying spike jump landing biomechanics in female adolescent volleyball athletes using video and verbal feedback. *The Journal of Strength & Conditioning Research,* 26, p. 1076-1084.

PATTON, M. Q. 1990. *Qualitative evaluation and research methods,* SAGE Publications, inc.

PAYTON, C. & BARTLETT, R. 2007. *Biomechanical evaluation of movement in sport and exercise: the British Association of Sport and Exercise Sciences guide,* Routledge.

PEARCE, A. 2005. Video analysis: Where do I start. *Sports Coach,* 28, p. 34-36.

PÉREZ, P., LLANA, S., BRIZUELA, G. & ENCARNACIÓN, A. 2009. Effects of three feedback conditions on aerobic swim speeds. *Journal of Sports Science & Medicine,* 8, p. 30.

PERSYN, U. Diagnosis of the movement and physical characteristics leading to feedback in the breaststroke. Competitive swimming Barcelona and beyond: Proceedings of the Pan Pacific International Swim Coaches Symposium, 1991. p. 30-62. PERSYN, U. & COLMAN, V. What are the best breaststrokers doing now. Applied Proceedings of the XVII International Symposium on Biomechanics in sports, 2005. p. 33-46.

PERSYN, U., COLMAN, V. & UNGERECHTS, B. 2005. Diagnosis & advice in the undulating strokes. *Coaches' Information Service-website: http://www.coachesinfo.com (online version 12/02/2007).*

PERSYN, U., COLMAN, V. & VAN TILBORGH, L. 1992. Movement analysis of flat and undulating breaststroke patterns. *Swimming Science VI (Biomechanics and medicine in swimming), London: E & FN SPON,* p. 75-80.

PERSYN, U., DE MAEYER, J. & VERVAECKE, H. 1975. Investigation of hydromechanics determinants of competitive swimming strokes. *Swimming II (International series on sport sciences), Baltimore (Maryland): University Park Press,* 2, p. 214222.

PERSYN, U., VERVAECKE, H. & VERHETSEL, D. 1983. Factors influencing stroke mechanics and speed in swimming the butterfly. *Biomechanics VIII-A & B,* p. 833-841.

PETERS, D. M. & O'DONOGHUE, P. 2013. *Performance analysis of sport IX,* Routledge.

PHILLIPS, E., FARROW, D., BALL, K. & HELMER, R. 2013. Harnessing and understanding feedback technology in applied settings. *Sports Medicine,* 43, p. 919-925.

PHILLIPS, S. 2015. *Fatigue in Sport and Exercise,* Routledge.

PIKE, D. 2008. *Investigating the Effects of Visually Augmented Video Feedback on Performance of an Ice.* School of Kinesiology, Lakehead University.

POLLOCK, C., JONES, I., JENKYN, T., IVANOVA, T. & GARLAND, S. 2012. Changes in kinematics and trunk electromyography during a 2000 m race simulation in elite female rowers. *Scandinavian Journal of Medicine & Science in Sports,* 22, p. 478-487.

PORTNEY, L. G. & WATKINS, M. P. 2000. *Foundations of clinical research: applications to practice,* Prentice Hall Upper Saddle River, NJ.

PRESTON, C. C. & COLMAN, A. M. 2000. Optimal number of response categories in rating scales: reliability, validity, discriminating power, and respondent preferences. *Acta Psychologica,* 104, p. 1-15.

PYNE, D. B., ANDERSON, M. E. & HOPKINS, W. G. 2006. Monitoring changes in lean mass of elite male and female swimmers. *International Journal of Sports Physiology and Performance,* 1, p. 14.

PYNE, D. B., TREWIN, C. B. & HOPKINS, W. G. 2004. Progression and variability of competitive performance of Olympic swimmers. *Journal of Sports Sciences,* 22, p. 613-620.

REMMONDS, P. & BARTLETT, R. 1981. Effects of finger separation. *Swimming Technique,* 18, p. 28-30.

RIBEIRO, L. F. P., LIMA, M. C. S. & GOBATTO, C. A. 2010. Changes in physiological and stroking parameters during interval swims at the slope of the d–t relationship.

Journal of Science and Medicine in Sport, 13, p. 141-145.

RICHMONDA, T., BUELLB, J., PFEILC, S. & CROWDERD, M. W. 2015. Evidence-based recommendations for maximizing competitive swimming performance. *Journal of Swimming Research,* 23, p. 1-14.

ROBERTSON, G., CALDWELL, G., HAMILL, J., KAMEN, G. & WHITTLESEY, S. 2013. *Research methods in biomechanics, 2E,* Human Kinetics.

ROBSON-ANSLEY, P. J., GLEESON, M. & ANSLEY, L. 2009. Fatigue management in the preparation of Olympic athletes. *Journal of Sports Sciences,* 27, p. 1409-1420.

ROYAL, K. A., FARROW, D., MUJIKA, I., HALSON, S. L., PYNE, D. & ABERNETHY, B. 2006. The effects of fatigue on decision making and shooting skill performance in water polo players. *Journal of Sports Sciences,* 24, p. 807-815.

RUCCI, J. A. & TOMPOROWSKI, P. D. 2010. Three types of kinematic feedback and the execution of the hang power clean. *The Journal of Strength & Conditioning Research,* 24, p. 771-778.

SALMONI, A. W., SCHMIDT, R. A. & WALTER, C. B. 1984. Knowledge of results and motor learning: a review and critical reappraisal. *Psychological Bulletin,* 95, p. 355.

SALO, A. I. & SCARBOROUGH, S. 2006. Athletics: Changes in technique within a sprint hurdle run. *Sports Biomechanics,* 5, p. 155-166.

SANDERS, R. 1996. Breaststroke technique variations among New Zealand Pan Pacific squad swimmers. *Swimming science VII. London: E & FN SPON,* p. 64-69.

SANDERS, R. H., CAPPAERT, J. M. & PEASE, D. L. 1998. Wave characteristics of Olympic breaststroke swimmers. *Journal of Applied Biomechanics,* 14, p. 40-51.

SARGENT, C., HALSON, S. & ROACH, G. D. 2014. Sleep or swim? Early-morning training severely restricts the amount of sleep obtained by elite swimmers. *European Journal of Sport Science,* 14, p. S310-S315.

SARRE, G., LEPERS, R. & VAN HOECKE, J. 2005. Stability of pedalling mechanics during a prolonged cycling exercise performed at different cadences. *Journal of Sports Sciences,* 23, p. 693-701.

SAWKA, M. N., BURKE, L. M., EICHNER, E. R., MAUGHAN, R. J., MONTAIN, S. J. & STACHENFELD, N. S. 2007. American College of Sports Medicine position stand. Exercise and fluid replacement. *Medicine and Science in Sports and Exercise,* 39, p. 377-390.

SCHLEIHAUF, R. 1979. A hydrodynamic analysis of swimming propulsion. *Swimming III,* p. 70-109.

SCHMIDT, R. A. & LEE, T. D. 1999. *Motor control and learning: A behavioural emphasis,* , Champaign, IL: Human Kinetics.

SCHNITZLER, C., SEIFERT, L. & CHOLLET, D. 2009. Variability of coordination parameters at 400-m front crawl swimming pace. *Journal of Sports Science & Medicine,* 8, p. 203.

SEIFERT, L., BOULESTEIX, L. & CHOLLET, D. 2004a. Effect of gender on the adaptation of arm coordination in front crawl. *International Journal of Sports Medicine,* 25, p. 217-223.

SEIFERT, L. & CHOLLET, D. 2005. A new index of flat breaststroke propulsion: A comparison of elite men and women. *Journal of Sports Sciences,* 23, p. 309-320.

SEIFERT, L., CHOLLET, D. & ALLARD, P. 2005. Arm coordination symmetry and breathing effect in front crawl. *Human Movement Science,* 24, p. 234-256.

SEIFERT, L., CHOLLET, D. & BARDY, B. 2004b. Effect of swimming velocity on arm coordination in the front crawl: a dynamic analysis. *Journal of Sports Sciences,* 22, 651-660.

SEIFERT, L., CHOLLET, D. & MUJIKA, I. 2011a. *World book of swimming: from science to performance,* Nova Science Publishers.

SEIFERT, L., CHOLLET, D., PAPPARADOPOULOS, C., GUERNIOU, Y. & BINET, G. 2006. Longitudinal evaluation of breaststroke spatial–temporal and coordinative parameters: Preparing of the 100-m breaststroke bronze medallists of the Athena 2004 Olympic Games. *Portuguese Journal of Sport Sciences,* 6, p. 260262.

SEIFERT, L., CHOLLET, D. & ROUARD, A. 2007. Swimming constraints and arm coordination. *Human Movement Science,* 26, p. 68-86.

SEIFERT, L., LEBLANC, H., HÉRAULT, R., KOMAR, J., BUTTON, C. & CHOLLET, D. 2011b.

Inter-individual variability in the upper–lower limb breaststroke coordination. *Human Movement Science,* 30, p. 550-565.

SEIFERT, L., TOUSSAINT, H., ALBERTY, M., SCHNITZLER, C. & CHOLLET, D. 2010. Arm coordination, power, and swim efficiency in national and regional front crawl swimmers. *Human Movement Science,* 29, p. 426-439.

SHARP, R., HACKNEY, A., CAIN, S. & NESS, R. 1988. The effect of shaving body hair on the physiological cost of freestyle swimming. *Journal of Swimming Research,* 4, p. 9-13.

SHIRREFFS, S. M. & MAUGHAN, R. J. 2006. The effect of alcohol on athletic performance. *Current Sports Medicine Reports,* 5, p. 192-196.

SIDNEY, M., DELHAYE, B., BAILLON, M. & PELAYO, P. 1999. Stroke frequency evolution during 100-m and 200-m events front crawl swimming. *Biomechanics and medicine in swimming VIII,* p. 71-75.

SLAWSON, S., JUSTHAM, L., WEST, A., CONWAY, P., CAINE, M. & HARRISON, R. 2008. Accelerometer profile recognition of swimming strokes (p17). *The Engineering of Sport 7.* Springer.

SMITH, A., ROBERTS, J., WALLACE, E., KONG, P. W., FORRESTER, S., MACKENZIE, S. & ROBERTSON, S. 2015. Golf Coaches' Perceptions of Key Technical Swing Parameters Compared to Biomechanical Literature. *International Journal of Sports Science and Coaching,* 10, 739-756.

SMITH, D. J. 2003. A framework for understanding the training process leading to elite performance. *Sports Medicine,* 33, p. 1103-1126.

SMITH, D. J., NORRIS, S. R. & HOGG, J. M. 2002. Performance evaluation of swimmers. *Sports Medicine,* 32, p. 539-554.

SMITH, R. E., SMOLL, F. L. & CUMMING, S. P. 2007. Effects of a motivational climate inntervention for coaches on young athletes' sport performance anxiety. *Journal of Sport & Exercise Psychology,* p. 39-59.

SOONS, B., SILVA, A., COLMAN, V. & PERSYN, U. 2003. Specific movement variables important for performance in different breaststroke styles. *Biomechanics and Medicine in Swimming IX,* p. 179-184.

SPENCER, M., FITZSIMONS, M., DAWSON, B., BISHOP, D. & GOODMAN, C. 2006. Reliability of a repeated-sprint test for field-hockey. *Journal of Science and Medicine in Sport,* 9, p. 181-184.

SPINKS, W. L. 1997. Sports research and the coach. *Sports Coach,* 19, p. 18-19.

STIRN, I., JARM, T., KAPUS, V. & STROJNIK, V. 2011. Evaluation of muscle fatigue during 100-m front crawl. *European Journal of Applied Physiology,* 111, p. 101-113.

STONE, M. H., STONE, M., SANDS, W. A. & SANDS, B. 2007. *Principles and practice of resistance training,* Human Kinetics.

STRZALA, M., KREZALEK, P., GLAB, G., KACA, M., OSTROWSKI, A., STANULA, A. & TYKA, A. K. 2013. Intra-cyclic phases of arm-leg movement and index of coordination in relation to sprint breaststroke swimming in young swimmers. *Journal of Sports Science & Medicine,* 12, p. 690.

STRZAŁA, M., KRĘŻĄŁEK, P., KUCIA-CZYSZCZOŃ, K., OSTROWSKI, A., STANULA, A., TYKA, A. K. & SAGALARA, A. 2014. Coordination and propulsion and nonpropulsion phases in 100 meters breaststroke swimming. *Acta of Bioengineering and Biomechanics,* 16, p. 83--89.

SUITO, H., IKEGAMI, Y., NUNOME, H., SANO, S., SHINKAI, H. & TSUJIMOTO, N. 2008. The effect of fatigue on the underwater arm stroke motion in the 100-m front crawl. *Journal of Applied Biomechanics,* 24, p. 316-324.

SUITO, H., TSUJIMOTO, N., SHINKAI, H., SANO, S., NUNOME, H. & IKEGAMI, Y. 2007. THE EFFECT OF FATIGUE ON THE

UNDERWATER ARM STROKE MOTION IN THE 100 M FRONT CRAWL. *Journal of Biomechanics,* 40, p. S772.

SULLIVAN, G. M. & FEINN, R. 2012. Using effect size-or why the P value is not enough. *Journal of Graduate Medical Education,* 4, p. 279-282.

SWART, J., LINDSAY, T. R., LAMBERT, M. I., BROWN, J. C. & NOAKES, T. D. 2012. Perceptual cues in the regulation of exercise performance–physical sensations of exercise and awareness of effort interact as separate cues. *British Journal of Sports Medicine,* 46, p. 42-48.

SWEETENHAM, B. & ATKINSON, J. 2003. *Championship swim training,* Human Kinetics.

SZABO, A. S. 2012. What is the meaning of sport coaching?(Expectations, possibilities, realities). *Sport Scientific & Practical Aspects,* 9, p. 39-44.

TAKAGI, H., SUGIMOTO, S., NISHIJIMA, N. & WILSON, B. 2004. Swimming: Differences in stroke phases, arm-leg coordination and velocity fluctuation due to event, gender and performance level in breaststroke. *Sports Biomechanics,* 3, p. 15-27.

TAN, F., POLGLAZE, T. & DAWSON, B. 2010. Reliability of an in-water repeated-sprint test for water polo. *International journal of sports physiology and performance,* 5, 117-120.

TAYLOR, K., CHAPMAN, D. W., CRONIN, J. B., NEWTON, M. J. & GILL, N. 2012. Fatigue monitoring in high performance sport: A survey of current trends. *Journal of Australian Strength and Conditioning,* 20, p. 12-23.

TELLA, V., TOCA-HERRERA, J., GALLACH, J., BENAVENT, J., GONZÁLEZ, L. & ARELLANO, R. 2008. Effect of fatigue on the intra-cycle acceleration in front crawl swimming: A time–frequency analysis. *Journal of Biomechanics,* 41, p. 86-92.

THOMAS, J. R., NELSON, J. K. & SILVERMAN, S. J. 2011. *Research Methods in Physical Activity,* Human Kinetics.

THOMPSON, A., BEZODIS, I. N. & JONES, R. L. 2009. An in-depth assessment of expert sprint coaches' technical knowledge. *Journal of Sports Sciences,* 27, p. 855-861.

THOMPSON, K., HALJAND, R. & MACLAREN, D. 2000. An analysis of selected kinematic variables in national and elite male and female 100-m and 200-m breaststroke swimmers. *Journal of Sports Sciences,* 18, p. 421-431.

THOMPSON, K. G., HALJAND, R. & LINDLEY, M. 2004. A Comparison of Selected Kinematic Variables Between Races in National to

Elite Male 200 m Breaststroke Swimmers. *Journal of Swimming Research,* 16, p. 6-10.

THOW, J. 2010. The Development and Validation of a Battery of Swimming Technique Measures Associated with Strength Deficits among Competitive Front-crawl Swimmers. The University of Edinburgh.

THOW, J. L., NAEMI, R. & SANDERS, R. H. 2012. Comparison of modes of feedback on glide performance in swimming. *Journal of sports sciences,* 30, 43-52.

TOUBEKIS, A. G., DOUDA, H. T. & TOKMAKIDIS, S. P. 2005. Influence of different rest intervals during active or passive recovery on repeated sprint swimming performance. *European Journal of Applied Physiology,* 93, p. 694-700.

TOURNY, C., CHOLLET, D., MICALLEF, J. & MACABIES, J. 1992. Comparative analysis of studies of speed variations within a breaststroke cycle. *Swimming science VI,* p. 161-166.

TOUSSAINT, H. 2011. Biomechanics of drag and propulsion in front crawl swimming. *World Book of Swimming: From Science to Performance,* p. 3-20.

TOUSSAINT, H. M. 2007. Strength power and technique of swimming performance: Science meets practice. *Leopold, W. Schwimmtrainer–Vereinigung V, Beucha, Deutschland,* p. 43-54.

TOUSSAINT, H. M. & BEEK, P. J. 1992. Biomechanics of competitive front crawl swimming. *Sports Medicine,* 13, p. 8-24.

TOUSSAINT, H. M., CAROL, A., KRANENBORG, H. & TRUIJENS, M. J. 2006. Effect of fatigue on stroking characteristics in an arms-only 100-m front-crawl race. *Medicine and Science in Sports and Exercise,* 38, p. 1635.

TREWIN, C. B., HOPKINS, W. G. & PYNE, D. B. 2004. Relationship between worldranking and Olympic performance of swimmers. *Journal of Sports Sciences,* 22, p. 339-345.

TURNER, A. P., SMITH, T. & COLEMAN, S. G. 2008. Use of an audio-paced incremental swimming test in young national-level swimmers. *International Journal of Sports Physiology and Performance,* 3, p. 68.

ULMER, H.-V. 1996. Concept of an extracellular regulation of muscular metabolic rate during heavy exercise in humans by psychophysiological feedback. *Experientia,* 52, p. 416-420.

VAN TILBORGH, L., WILLEMS, E. & PERSYN, U. 1988. Estimation of breaststroke propulsion and resistance resultant impulses from film analysis. *Swimming Science V (International series on sport sciences 18), Champaign (Ilinois): Human Kinetics*, p. 67-72.

VARGAS-TONSING, T. 2007. Coaches' preferences for continuing coaching education. *International Journal of Sports Science and Coaching,* 2, p. 25-35.

VEZOS, N., GOURGOULIS, V., AGGELOUSSIS, N., KASIMATIS, P., CHRISTOFORIDIS, C. & MAVROMATIS, G. 2007. Underwater stroke kinematics during breathing and breath-holding front crawl swimming. *Journal of Sports Science & Medicine,* 6, P. 58.

VILAS-BOAS, J. 1996. Speed fluctuations and energy cost of different breaststroke techniques. *Biomechanics and Medicine in Swimming,* 7, P. 167-171.

VILAS-BOAS, J. P. & SANTOS, P. 1994. Comparison of swimming economy in three breaststroke techniques. *Medicine and Science in Aquatic Sports*, P. 48-54.

VØLLESTAD, N. K. 1997. Measurement of human muscle fatigue. *Journal of Neuroscience Methods,* 74, p. 219-227.

VORONTSOV, A. & RUMYANTSEV, V. 2000. Propulsive forces in swimming. *Biomechanics in sport–Performance, Enhancement and Injury Prevention*, p. 205231.

WAKAYOSHI, K., D'ACQUISTO, J., CAPPAERT, J. & TROUP, J. 1996. Relationship between metabolic parameters and stroking technique characteristics in front crawl. *Biomechanics and Medicine in Swimming VII*, p. 152-158.

WALLACE, L. K., SLATTERY, K. M. & COUTTS, A. J. 2009. The ecological validity and application of the session-RPE method for quantifying training loads in swimming. *The Journal of Strength & Conditioning Research,* 23, p. 33-38.

WANG, J. R. & PARAMESWARAN, N. Survey of sports video analysis: research issues and applications. Proceedings of the Pan-Sydney area workshop on Visual information processing, 2004. Australian Computer Society, Inc., p. 87-90.

WATERS, A., LAY, B., TIDMAN, S., BENJANUVATRA, N., WATERS, A., LAY, B., TIDMAN, S., BENJANUVATRA, N. & IN, B. M. Visual search behaviour and information extraction differences between high-level and developing swimming coaches. Proceedings of the Xllth lnternational Symposium for Biomechanics and Medicine in Swimming, 2014 Australia. p. 343-348.

WESTERBLAD, H., BRUTON, J. D. & KATZ, A. 2010. Skeletal muscle: energy metabolism, fiber types, fatigue and adaptability. *Experimental Cell Research,* 316, p. 30933099.

WILKIE, D. & JUBA, K. 1994. *The handbook of swimming,* Pelham Books.

WILLIAMS, A. M. & HODGES, N. J. 2005. Practice, instruction and skill acquisition in soccer: Challenging tradition. *Journal of Sports Sciences,* 23, p. 637-650.

WILLIAMS, S. J. & KENDALL, L. 2007. Perceptions of elite coaches and sports scientists of the research needs for elite coaching practice. *Journal of Sports Sciences,* 25, p. 1577-1586.

WILSON, B. D. 2008. Development in video technology for coaching. *Sports Technology,* 1, p. 34-40.

WREN, T. A., DO, K. P., HARA, R. & RETHLEFSEN, S. A. 2008. Use of a patella marker to improve tracking of dynamic hip rotation range of motion. *Gait & Posture,* 27, p. 530-534.

WULF, G. 2007. Attentional focus and motor learning: A review of 10 years of research. *E-journal Bewegung und Training,* 1, p. 1-11.

WULF, G. & PRINZ, W. 2001. Directing attention to movement effects enhances learning: A review. *Psychonomic Bulletin & Review,* 8, p. 648-660.

WULF, G. & SHEA, C. H. 2004. Understanding the role of augmented feedback. *Skill Acquisition in Sport: Research, Theory and Practice,* p. 121-144.

ZAIDI, H., TAIAR, R., FOHANNO, S. & POLIDORI, G. 2008. Analysis of the effect of swimmer's head position on swimming performance using computational fluid dynamics. *Journal of Biomechanics,* 41, p. 1350-1358.

ZAMPARO, P., GATTA, G., PENDERGAST, D. & CAPELLI, C. 2009. Active and passive drag:

the role of trunk incline. *European Journal of Applied Physiology,* 106, p. 195205.

ZATOŃ, K. & SZCZEPAN, S. 2012. The Effect of Immediate Verbal Feedback on the Efficiency and the Effectiveness of Swimming. *Baltic Journal of Health and Physical Activity,* 4, p. 91-103.

Made in the USA
Monee, IL
12 November 2026

275e12d5-42f4-4670-a7a6-ee706e5b12f0R02